the UNIVERSE
and how to see it

the UNIVERSE
and how to see it

**A practical guide to viewing and
understanding the night sky**

Giles Sparrow

MARSHALL PUBLISHING • LONDON

Contents

A MARSHALL EDITION

Conceived, edited and designed
by Marshall Editions Ltd
The Lawns
74 Shepherd's Bush Green
Shepherd's Bush
London
W12 8QE

First published in Great Britain in 2001
by Marshall Editions Ltd

Project Editor	Dan Green
Sub-Editors	Sue Harper
	Peter Adams
	Anna Southgate
	Nicola Munro
Managing Art Editor	Helen Spencer
Designers	Joyce Mason,
	Anne Fisher
	Paul Montague
	Philip Letsu
Visualizer	Iain Stuart
Picture Editor	Antonella Mauro
Proofreader	Peter Kirkham
Indexer	Helen Smith
Production Controller	Anna Pauletti
Editorial Coordinator	Gillian Thompson
Art Director	Dave Goodman
Editorial Director	Ellen Dupont

Originated in Singapore by Master Image
Printed and bound in Portugal by Printer
Portugesa.
1 3 5 7 9 10 8 6 4 2

Jacket Credits:
Front, Main image Bruce Coleman; t Bruce
Coleman,
c Anglo-Australian Observatory/David Malin,
b NASA/David Seal. Front flap, James Gitlin
(STScI). Back, Bruce Coleman.

Getting started
10-29

This section introduces the sky.
It explains how the constellations
change depending on the time of
year and your location on Earth,
and offers hints on how to navigate
your way around the heavens.

The Solar System
30-79

An introduction to what is known
about the planets and other bodies
of our Solar System. This section
also tells you what to look for and
how to locate these ever-changing
objects in the night sky.

How to use this book

The Universe and how to see it offers a unique approach to the wonders of the cosmos. Each of the book's double-page spreads explores a particular aspect of the Universe, and details how to find them in the night sky.

Title and page number
Clear titles and prominent page numbering aid easy navigation within the book, and allow you to cross-refer between the foldout Constellation guides (pp. 12-13) and the pages where the individual constellations are covered in detail.

Section headings
The book is divided into five sections, each dealing with a realm of the cosmos. A glance at this bar will tell you where you are in the Universe.

Introduction
A short overview of the subject matter covered on the pages.

Main text
This section provides a concise description of the basics, but also includes the most recent discoveries that have altered our traditional understanding of the subject.

Stunning images
We have gathered together the most up-to-the-minute, awe-inspiring images from the world's major observatories, orbiting telescopes, spaceprobes and leading space artists.

Cross-reference bar
Jump to other pages in the book that have related topics.

Planet and star profiles
To keep the text free of too many figures we have put all the essential data about planets and other astronomical objects into easy-to-read tables.

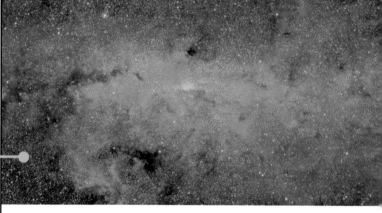

STARS AND THE GALAXY

136 The centre of our galaxy

Deep in Sagittarius, 26,000 light-years away, lies the most violent place in our cosmic neighbourhood – home to enormous stars, fountains of antimatter and a giant black hole weighing more than two million Suns.

In the centre of the Milky Way, stars are closely packed together, and may even collide at times, merging to form stellar giants that break all the rules of normal stellar evolution. Most of the stars in the central nucleus are old and red. They formed early in the galaxy's history and have few of the heavy elements that help to speed up the life cycles of younger stars. These stellar fossils are next door to much younger neighbours – the gas and dust clouds in the nucleus are production lines for clusters of bright, massive stars.

The galaxy's core is hidden by a ring of dense clouds surrounding a central cavity ten light-years across that completely block visible light. By using infrared light, however, astronomers can see stars at the very heart of the Milky Way. Their movements show that they are being affected by an object with 2.6 million times the mass of the Sun, crammed into a space just one-thirtieth of a light-year across. With such incredible density, this object can only be a giant black hole.

Today, this black hole is asleep. The area within its gravitational reach has been swept clear of material, and the giant stars around it keep their distance. A little dust still falls steadily inward, heating up and emitting radio waves. But things were not always so quiet. Thousands of light-years above the nucleus lies a huge fountain of antimatter, glowing with high-energy gamma rays. The antimatter was probably blasted out of the nucleus the last time that an ill-fated star drifted into the black hole's grasp, and the sleeping giant awoke.

Fire in the darkness This image shows how the central Milky Way would appear if our eyesight was shifted into the infrared, allowing us to peer through the dark dust clouds that surround the nucleus. Every single speck of light is an individual star, and towering filaments of gas rise high above the horizontal plane of the Milky Way, following the lines of the galaxy's magnetic field.

See also: Star distances 84 · Star colours 88 · Stellar evolution 92 · Red & brown dwarfs 102 · Stellar heavyweights 104

PLANET PROFILE

Visual magnitude (avg.)	-2.7	Equatorial diameter	142,984 km	Liquid hydrogen
Distance from Sun (avg.)	778.3 million km		88,846 miles	
	483.6 million miles	Mass (Earth = 1)	318	Rocky core
	5.203 AU	Gravity at equator (Earth = 1)	2.6	
Length of year	11.86 Earth years	Cloud-top temperature	-110 °C	Liquid metallic hydrogen – individual charged atoms of hydrogen
Rotation period	9.93 Earth hours		-160 °F	

See also: The home system 32 · Genesis 34 · The Galilean moons 66 · Probes to the planets 202

Constellation information

The caption to the star map explains where the constellation can be seen, and its most interesting features. The numbered box next to the title allows you to cross-refer to the foldout Constellation guides (pp. 12-13) at the front of the book.

Star stories

These boxes give you the legends and stories behind the stars, planets, and constellations. Cultures are included from all over the world from ancient Greece to Australian Aborigines.

Astrophotos

Stunning astrophotos, many taken by amateurs, highlight the most visible and striking objects in the constellation.

Planet locater diagrams

This unique device devised by the author allows you to work out the position and visibility of planets at any point over the next ten years or more.

Objects of interest

The stars, nebulae, and galaxies in the constellation that demonstrate the subject matter on the facing page, are clearly labelled on the main star map, and linked to the descriptions surrounding it.

Object descriptions

Individual objects are described in more detail, with information on what they look like, and whether they are visible with the naked eye, binoculars, or a telescope. Wherever possible, we have picked out naked-eye and binocular objects.

Constellation map

We have carefully selected constellations that show the celestial objects in discussion. Specially commissioned, these star maps are designed to be as realistic as possible. Where a constellation is visible only from one hemisphere, please refer to the Celestial object index (pp. 216-217) for a list of alternative objects to view.

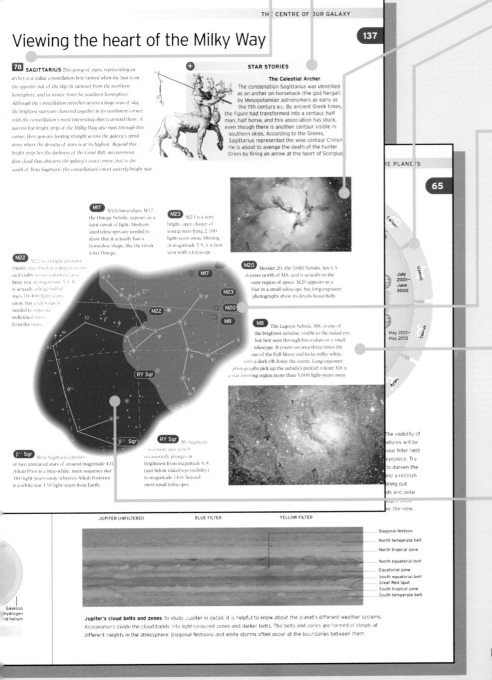

THE CENTRE OF OUR GALAXY

Viewing the heart of the Milky Way

137

78 SAGITTARIUS *This group of stars, representing an archer, is a zodiac constellation best viewed when the Sun is on the opposite side of the sky: in summer from the northern hemisphere, and in winter from the southern hemisphere. Although the constellation stretches across a large area of sky, the brightest stars are clustered together in its northwest corner, with the constellation's most interesting objects around them. A narrow but bright strip of the Milky Way also runs through this corner. Here you are looking straight across the galaxy's spiral arms where the density of stars is at its highest. Beyond this bright strip lies the darkness of the Great Rift, an enormous dust cloud that obscures the galaxy's exact centre, just to the south of Beta Sagittarii, the constellation's most easterly bright star.*

STAR STORIES

The Celestial Archer

The constellation Sagittarius was identified as an archer on horseback (the god Nergal) by Mesopotamian astronomers as early as the 11th century B.C. By ancient Greek times, the figure had transformed into a centaur, half man, half horse, and this association has stuck, even though there is another centaur visible in southern skies. According to the Greeks, Sagittarius represented the wise centaur Chiron. He is about to avenge the death of the hunter Orion by firing an arrow at the heart of Scorpius.

M17 With binoculars, M17, the Omega Nebula, appears as a faint streak of light. Medium-sized telescopes are needed to show that it actually has a horseshoe shape, like the Greek letter Omega.

M23 M23 is a very bright, open cluster of young stars lying 2,100 light-years away. Shining at magnitude 5.9, it is best seen with a telescope.

M22 M22 is a bright globular cluster, one-third of a degree across and visible to the naked eye as a fuzzy star at magnitude 5.1. It is actually a huge ball of stars 10,400 light-years away, but a telescope is needed to separate individual stars from the mass.

M20 Messier 20, the Trifid Nebula, lies 1.5 degrees north of M8, and is actually in the same region of space. M20 appears as a blur in a small telescope, but long-exposure photographs show its details beautifully.

M8 The Lagoon Nebula, M8, is one of the brightest nebulae, visible to the naked eye, but best seen through binoculars or a small telescope. It covers an area three times the size of the Full Moon and looks milky-white, with a dark rift down the centre. Long-exposure photographs pick up the nebula's pinkish colour. M8 is a star forming region more than 5,000 light-years away.

β¹² Sgr Beta Sagittarii consists of two unrelated stars of around magnitude 4.0. Arkab Prior is a blue-white, main sequence star 380 light-years away, whereas Arkab Posterior is a white star 139 light-years from Earth.

RY Sgr RY Sagittarii is a study star which occasionally plunges in brightness from magnitude 6.8 (just below naked-eye visibility) to magnitude 14.0, beyond most small telescopes.

THE PLANETS

65

Cancer

July 2001– June 2002

Gemini

May 2012– May 2013

Taurus

Aries

The visibility of ...atures will be ...our filter held ...eyepiece. Try ...o darken the ...nd a reddish ...ring out ...ds and polar ...ges show ...ve the view.

JUPITER UNFILTERED   BLUE FILTER   YELLOW FILTER

Diagonal festoon
North temperate belt
North tropical zone
North equatorial belt
Equatorial zone
South equatorial belt
Great Red Spot
South tropical zone
South temperate belt

Jupiter's cloud belts and zones To study Jupiter in detail, it is helpful to know about the planet's different weather systems. Astronomers divide the cloud bands into light-coloured zones and darker belts. The belts and zones are formed of clouds at different heights in the atmosphere. Diagonal festoons and white storms often occur at the boundaries between them.

Gaseous hydrogen and helium

Constellation guides

Keep these pages open as you read the book to help you locate individual constellations and see how they relate to each other.

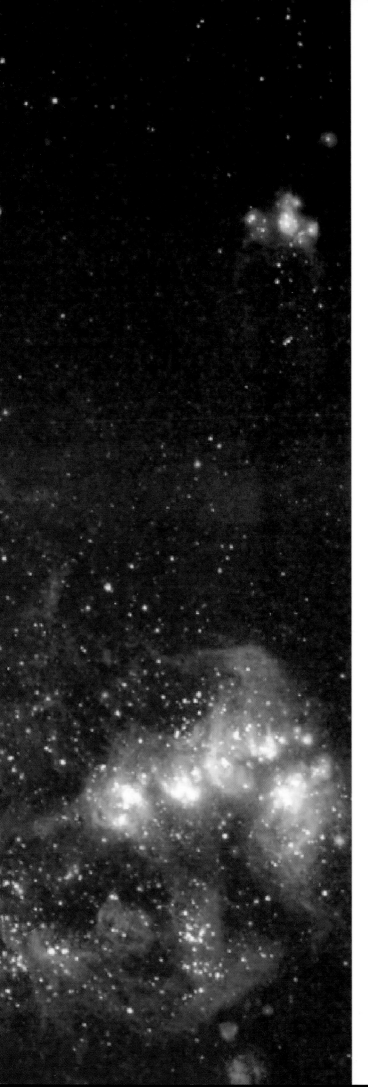

Introduction

Although I've been fascinated by the stars for longer than I can recall, I can still remember the simple question that turned astronomy into a lifelong passion: why, I wondered, was the Moon following me? As my parents' car sped along on a clear winter's night, we passed buildings, trees and fields, leaving them all behind – but the Moon kept constant pace with us.

Today, of course, I know it's all a matter of parallax and perspective – the same effect that lets us turn the separate images from our eyes into a three-dimensional view of the world. Simply put, because the Moon is a quarter of a million miles beyond even the highest clouds, our point of view of it is almost identical from anywhere on Earth.

A quarter of a million miles... It still sometimes gives me a jolt to think that the ball of rock apparently just above the clouds is in fact so distant, held in orbit around Earth by the threads of the same invisible force that makes my pen fall to the ground when it rolls off the table. And, to borrow a phrase, "that's just peanuts to space".

This book is all about such shifts in perspective. By combining the description of how the Universe works with examples of where to see it in action, I hope to give seasoned stargazers more insight into their favourite observing targets, and encourage armchair astronomers to step outside and see the reality behind the beautiful photographs and paintings. But above all, this book is written for anyone who has ever looked up on a starry night and wanted to know more.

Giles Sparrow

June 2001

14 The changing sky

The night sky is a wondrous, constantly changing spectacle. Binoculars and telescopes can reveal that it contains other worlds, countless stars, delicate nebulae, and distant galaxies. But the first step is simply to look up on a clear, dark night.

Many stars appear in distinctive patterns – the constellations. There are 88 constellations in all, though many are faint and indistinct. The bright ones tend to cluster around a band of faint light that runs across the sky – the Milky Way. Binoculars will show that the Milky Way is made up of millions of individual stars. Stars cluster together in this direction because we are looking across the flattened disc of our own galaxy, a system containing billions of stars.

On a dark night, there seem to be millions of stars, but in fact there are fewer than 3,000 stars visible with the naked eye at one time. The brightness of stars is defined by a number called magnitude – the lower the magnitude, the brighter the star. The very brightest stars, and some objects within the Solar System, have negative magnitudes. Sirius is the brightest star in the sky at magnitude -1.4, and the faintest stars visible to the naked eye shine at magnitude 6.0.

Individual stars and planets are not the only sources of light in the heavens – the sky is scattered with star clusters, nebulae (gas clouds) and distant galaxies. These are usually named according to their number in one catalogue or another – nearly all the brightest objects appear in the Messier catalogue or New General Catalogue, and are given M or NGC numbers. Unlike stars, which are so distant that they appear as points of light, clusters, nebulae and galaxies appear as spread-out glows.

Astronomers use several different units to measure the vast distances of interplanetary, interstellar and intergalactic space. In the pages that follow, we shall use just two. Within the Solar System, the basic measurement unit is the Astronomical Unit (AU) – simply the average distance from Earth to the Sun, or 150 million km (93 million miles). Between the stars and beyond the galaxy, distances are best measured in light-years – the distance travelled by light, the fastest thing in the Universe, in one Earth year. This is roughly 9.5 trillion km (5.8 million million miles).

Star trails Watch for more than a few minutes and you will realize that the stars are not static. They are "crawling" across the sky from east to west reaching their highest point as they cross the line from north to south. As some constellations set in the west, new ones rise in the east. This movement is caused by Earth's daily rotation – the stars take 23 hours and 56 minutes to return to their original positions.

See also: The celestial sphere 16 · Navigating the skies 18 · The home system 32 · The starry sky 82 · Star distances 86

Getting started

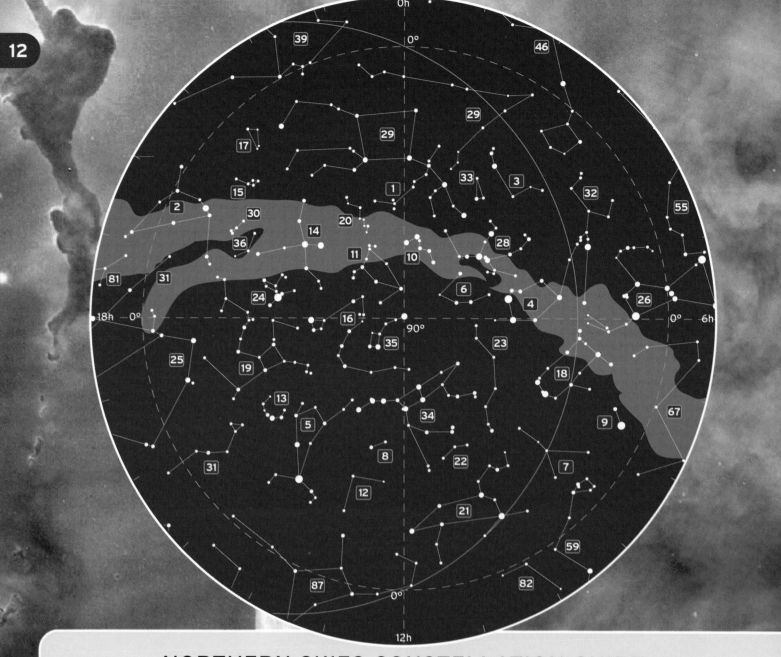

NORTHERN SKIES CONSTELLATION GUIDE
(Featured constellations are in **bold** type with page references)

1 **Andromeda 143**

2 Aquila

3 **Aries 145**

4 **Auriga 129**

5 **Boötes 99**

6 Camelopardalis

7 **Cancer 131**

8 **Canes Venatici 153**

9 Canis Minor

10 **Cassiopeia 97**

11 **Cepheus 115**

12 **Coma Berenices 155**

13 **Corona Borealis 125**

14 **Cygnus 127**

15 Delphinus

16 **Draco 125**

17 Equuleus

18 **Gemini 93**

19 **Hercules 133**

20 Lacerta

21 **Leo 91**

22 Leo Minor

23 Lynx

24 **Lyra 107**

25 **Ophiuchus 85**

26 **Orion 95**

27 **Pegasus 117**

28 **Perseus 109**

29 **Pisces 17**

30 Sagitta

31 **Serpens 85**

32 **Taurus 121**

33 **Triangulum 145**

34 **Ursa Major 83**

35 **Ursa Minor 15**

36 **Vulpecula 117**

SYMBOLS USED ON CONSTELLATION MAPS

Stars

Magnitude below 0.0

Magnitude 0.1–1.0

Magnitude 1.1–2.0

Magnitude 2.1–3.0

Magnitude 3.1–4.0

Magnitude 4.1–5.0

Magnitude 5.1–6.0

Deep-sky objects

Nebula

Open cluster

Globular cluster

Galaxy

Telescopic object

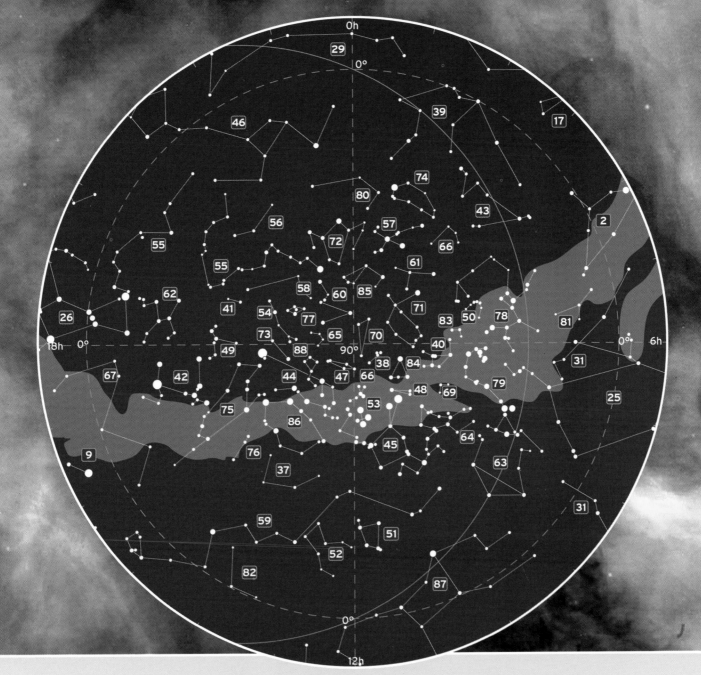

13

SOUTHERN SKIES CONSTELLATION GUIDE
(Featured constellations are in **bold** type with page references)

37 Antlia
38 Apus
39 Aquarius
40 Ara
41 Caelum
42 **Canis Major 89**
43 Capricornus
44 **Carina 105**
45 **Centaurus 87**
46 **Cetus 157**
47 Chamaeleon
48 Circinus
49 Columba

50 Corona Australis
51 Corvus
52 Crater
53 **Crux 111**
54 **Dorado 141**
55 **Eridanus 119**
56 **Fornax 149**
57 Grus
58 Horologium
59 Hydra
60 Hydrus
61 Indus
62 **Lepus 103**

63 Libra
64 Lupus
65 Mensa
66 Microscopium
67 Monoceros
68 Musca
69 Norma
70 **Octans 16**
71 Pavo
72 Phoenix
73 **Pictor 101**
74 Piscis Austrinus
75 **Puppis 135**

76 Pyxis
77 Reticulum
78 **Sagittarius 137**
79 **Scorpius 113**
80 **Sculptor 159**
81 Scutum
82 Sextans
83 Telescopium
84 Triangulum Australe
85 **Tucana 147**
86 **Vela 123**
87 **Virgo 151**
88 Volans

Homeworld As night falls over the California desert, Joshua trees stand out starkly against the horizon. The sky above the sunset slowly leaches of its colours, until just red remains. On the opposite side of the sky, Earth's shadow is rising, bringing with it a curtain of deep blue, and the promise of a clear night...

The celestial poles

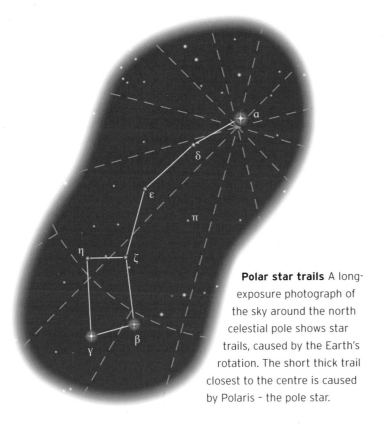

Polar star trails A long-exposure photograph of the sky around the north celestial pole shows star trails, caused by the Earth's rotation. The short thick trail closest to the centre is caused by Polaris - the pole star.

35 **URSA MINOR** *The north celestial pole, the point in the sky directly above Earth's north pole, lies in the constellation of Ursa Minor, the Little Bear. The constellation is similar in shape to Ursa Major and its brightest star lies within a Full Moon's breadth of the pole itself. This star, Polaris, is labelled Alpha, and is a 2nd-magnitude star. The other stars demonstrate the magnitude scale – Gamma is a 3rd-magnitude star and Epsilon and Eta are 4th and 5th magnitudes.*

70 **OCTANS** *In contrast with the north celestial pole, the sky's south pole lies in a relatively barren area of the sky – the constellation of Octans, the Octant – named after a navigational instrument. The nearest naked-eye star to the pole itself is Sigma Octantis, a magnitude 5.4 star.*

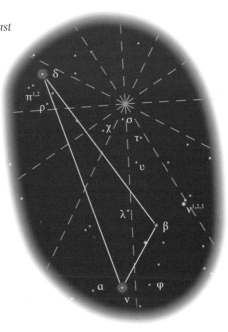

Stars of the southern hemisphere The southern hemisphere has no bright pole star, so there is a gap near the pole itself. As can clearly be seen, the farther stars get from the celestial pole, the faster their movement across the sky. The light smudges at top and top right are, respectively, the Large and Small Magellanic Clouds– small galaxies that orbit our own Milky Way Galaxy.

The celestial sphere

16

From Earth, the sky appears to form a vast sphere upon which the Sun, Moon and planets move. Although purely imaginary, this model of the Universe is very useful to astronomers for calculating the positions of objects in the sky.

From an observer's point of view, the celestial sphere seems to spin around Earth, pivoted at the north and south celestial poles in Ursa Minor and Octans – but it is Earth that rotates.

The sphere can be divided into small segments – 360 degrees – in order to define the position of any object in the sky. For greater accuracy, degrees are divided into minutes and seconds of arc, where one minute is $\frac{1}{60}$th degree, and one second is $\frac{1}{60}$th minute. Any object's position can be found by measuring its altitude above Earth's horizon, and its azimuth – the angle separating it, clockwise, from due north. This alt-azimuth system of coordinates is simple, but limited as Earth's rotation means that objects constantly change azimuth and altitude.

Far more useful is the equatorial coordinate system, which measures star positions relative to the celestial equator – the line midway between the north and south celestial poles. Here, the declination of a star is its angle above or below this line, measured in degrees, where the north celestial pole is +90 degrees, and the south celestial pole is -90 degrees.

For east-west positions, astronomers measure an object's position along a circle around the sky parallel to the celestial equator. To do this they have to designate a north-south line through the sky – a sort of celestial Greenwich Meridian – marking the start and end of such measurements.

The celestial sphere is defined with its poles fixed over Earth's poles, but

Earth orbits the Sun tilted at an angle of 23.5 degrees from vertical. So, in one year, the Sun moves from the northern to the southern hemisphere and back, on a path known as the ecliptic. In their search for a meaningful reference

point for east-west measurements, astronomers hit on the point where the Sun moves into the northern half of the sky, called the First Point of Aries.

Celestial longitudes, called right ascension or R.A., are not measured in degrees, but in hours, minutes and seconds of time. An object's R.A. is the time between the First Point of Aries – currently in Pisces – reaching its highest daily point in the sky, and the object doing the same thing.

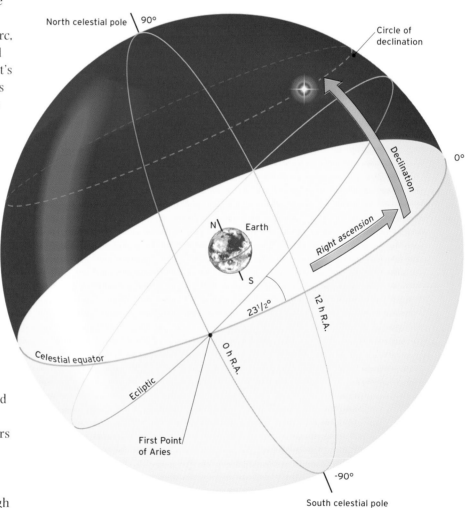

Ancient wisdom The celestial sphere harks back to a time when astronomers believed that a hollow sphere of fixed stars rotated around Earth. This model is used today in the equatorial coordinate system. Declination is the celestial latitude of an object in the sky, and right ascension is the star's longitude, measured anticlockwise from the First Point of Aries.

See also: The changing sky 14 · Navigating the skies 18 · Naming the stars 20 · The starry sky 82 · Star distances 86

Precession in Pisces

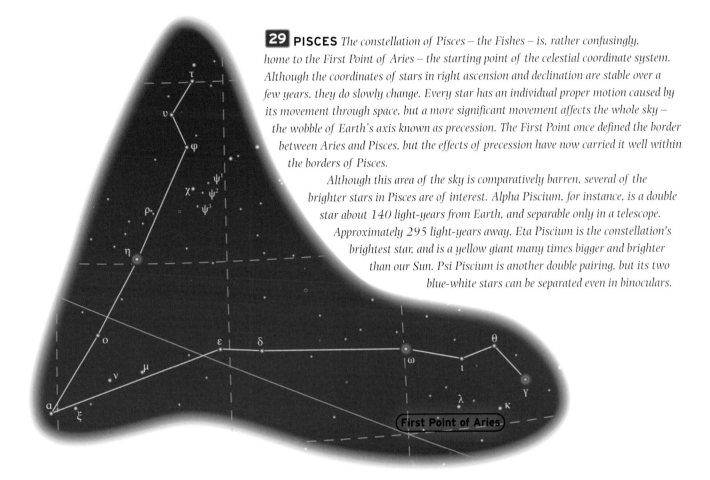

29 PISCES *The constellation of Pisces – the Fishes – is, rather confusingly, home to the First Point of Aries – the starting point of the celestial coordinate system. Although the coordinates of stars in right ascension and declination are stable over a few years, they do slowly change. Every star has an individual proper motion caused by its movement through space, but a more significant movement affects the whole sky – the wobble of Earth's axis known as precession. The First Point once defined the border between Aries and Pisces, but the effects of precession have now carried it well within the borders of Pisces.*

Although this area of the sky is comparatively barren, several of the brighter stars in Pisces are of interest. Alpha Piscium, for instance, is a double star about 140 light-years from Earth, and separable only in a telescope. Approximately 295 light-years away, Eta Piscium is the constellation's brightest star, and is a yellow giant many times bigger and brighter than our Sun. Psi Piscium is another double pairing, but its two blue-white stars can be separated even in binoculars.

First Point of Aries

Precession of the equinoxes

Precession is caused by the Sun and Moon's gravity pulling on the bulge around Earth's equator. These forces cause our planet to wobble slowly like a spinning top or gyroscope, describing a circle in the sky over 25,800 years. As well as affecting the First Point of Aries, precession affects the position of the celestial poles as Earth's axis points at different parts of the sky. Currently Polaris is the closest bright star to the north celestial pole, but 5,000 years ago, it was Thuban (Alpha Draconis). In A.D. 7500, Alderamin (Alpha Cephei) will be the pole star, and in A.D. 14000, it will be Vega (Alpha Lyrae).

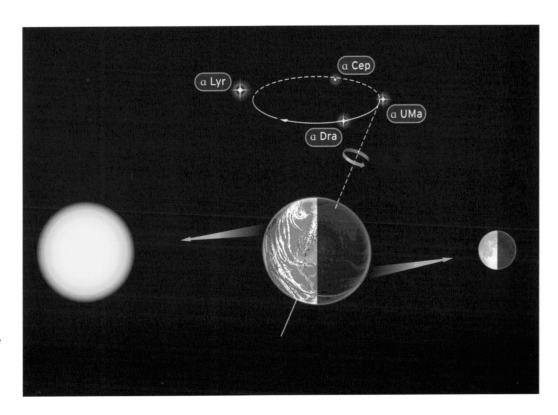

Navigating the skies

From any point on the surface of Earth, we can see only half of the celestial sphere. This means that constellations can be classed as either southern or northern hemisphere – they are mostly visible from only one hemisphere on Earth.

Not all stars and other celestial objects are visible to all observers everywhere. This is because Earth is so large that its horizon completely blocks off one side of the sky from our view. Instead, what we see can be thought of as the inside of a large, round mixing bowl. The edges of this hemisphere describe a circle – the horizon – that is separated by 90 degrees from the point directly overhead – the zenith.

To an observer standing at either of Earth's poles, the celestial pole is at the zenith. Its altitude above the horizon is 90 degrees, so the celestial equator runs around the horizon. The sky visible to the observer consists of an entire celestial hemisphere, and nothing else. These two viewpoints – from north and south poles – are shown on the Constellation guides (see pp. 12–13).

However, most observers are not at these extreme positions but somewhere in between Earth's equator and either pole, so they see a mix of stars from both the northern and southern hemispheres. The visible celestial pole loses altitude as you travel toward the equator, until at the equator it is on the horizon – altitude 0 degrees. In fact, the altitude of the celestial pole is always equal to an observer's latitude – a fact that proved invaluable to seafarers navigating by the stars.

If the celestial pole is lower in the sky, then the celestial equator is higher. At Earth's equator, the celestial equator runs from east to west, passing directly overhead, but this is a special case. In general, the celestial equator is at its highest altitude as it crosses an imaginary line drawn from the zenith away from the pole – the meridian. At this point, the equator's altitude is always 90 degrees minus the observer's latitude, so an observer at latitude 50 degrees sees the equator cross the meridian at altitude 40 degrees.

Therefore, you can see stars in the opposite celestial hemisphere from any location, provided they are not farther south – or north – than 90 degrees minus your latitude. These stars will be best seen as they cross the meridian, but the constellation pattern will be flipped upside down compared with its appearance on the opposite side of the equator. To avoid confusion, star maps traditionally – and in this book – treat north as up, regardless of which hemisphere a constellation is in.

Similarly, the stars within a certain distance of the celestial pole – a circle with radius equivalent to the observer's latitude – will never set. These circumpolar constellations vary with latitude, but for most northern hemisphere observers they include Ursa Minor, Ursa Major, Draco, Cepheus and Cassiopeia. These constellations will circle the north celestial pole every 23 hours, 56 minutes, dipping low on the northern horizon, but never vanishing completely.

For observers in the southern hemisphere, there is no equivalent of the pole star and the circumpolar stars are largely a scattering of faint constellations around the south celestial pole. However, for most observers they also include Crux, the Large and Small Magellanic Clouds, Carina and parts of Centaurus.

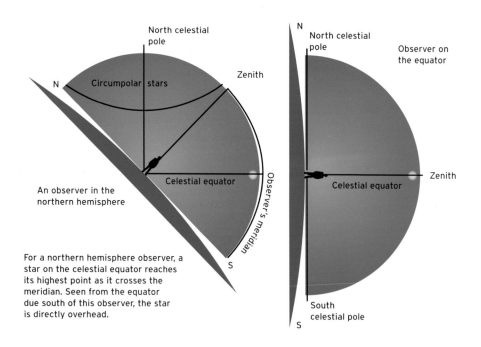

For a northern hemisphere observer, a star on the celestial equator reaches its highest point as it crosses the meridian. Seen from the equator due south of this observer, the star is directly overhead.

Your sky is not my sky As the latitude of the observer changes, the section of the celestial sphere that is visible to them also changes. The star that appears to be at 45 degrees above the horizon for an observer at mid-northern latitudes appears directly overhead - at 90 degrees - to an observer near the equator.

See also: The changing sky 14 · The celestial sphere 16 · Naming the stars 20 · The starry sky 82

Signposts of the northern sky

Northern wayfinders *The skies of the northern hemisphere contain two bright groups of stars long established as celestial signposts. Ursa Major, the Great Bear, can be used to locate most of the other northern circumpolar stars, while the Summer Triangle dominates the southern aspect of the sky on nights from July to September, and also contains useful pointers.*

5 BOÖTES
8 CANES VENATICI
10 CASSIOPEIA
11 CEPHEUS

13 CORONA BOREALIS
16 DRACO

31 SERPENS
34 URSA MAJOR
35 URSA MINOR

Ursa Major group The brightest seven stars of Ursa Major form the Plough and are the key to the northern polar stars. The end stars of the Plough, Dubhe and Merak, point to the Pole Star Polaris (Alpha Ursae Minoris), and through this toward Cassiopeia and Pegasus. A curved line from the Plough's handle leads to Arcturus (Alpha Boötis), and through this to Virgo, while a line across the top of the Plough points to Auriga, the Charioteer.

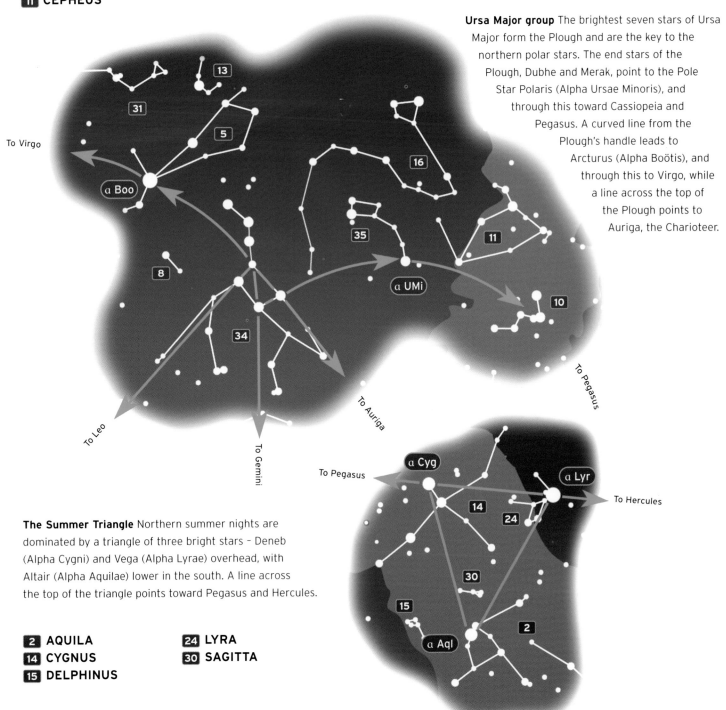

The Summer Triangle Northern summer nights are dominated by a triangle of three bright stars – Deneb (Alpha Cygni) and Vega (Alpha Lyrae) overhead, with Altair (Alpha Aquilae) lower in the south. A line across the top of the triangle points toward Pegasus and Hercules.

2 AQUILA
14 CYGNUS
15 DELPHINUS

24 LYRA
30 SAGITTA

Naming the stars

Even with the unaided eye, thousands of stars – as well as numerous star clusters, nebulae and galaxies – can be seen. Astronomers have developed many naming conventions to keep track of all the different objects.

From ancient times, astronomers divided the stars into patterns that represented animals, familiar objects and their mythological heroes – the constellations. Today's 88 constellations are based on the 48 designated by the Greek-Egyptian astronomer Ptolemy in the 2nd century, with additions by various astronomers, between the 16th and 18th centuries, to cover uncharted areas such as the far southern skies. The constellations are now known internationally by Latin names, with standard three-letter abbreviations (see pp. 204-205).

The original constellations were viewed as patterns made from individual stars, but telescopes showed that the naked-eye stars were just the brightest in any area of sky. This presented the problem of where to fit in any newly discovered objects. Finally, in 1922 the International Astronomical Union redefined the constellations as areas of sky with precise borders. This means that every new object can now be accommodated within a constellation.

Many of the brightest stars have individual names, often derived from Arabic, but the most important naming system is the Bayer-Flamsteed classification. In 1603 the German astronomer Johann Bayer suggested that the brightest stars in a constellation should be given Greek letters to indicate their brightness – α (alpha) indicating the brightest star in a constellation, β (beta), the second brightest, and so on.

Bayer's innovation is useful, but can only be used for a constellation's 24 brightest stars. A century later the British Astronomer Royal John Flamsteed devised a better scheme – numbering the stars in a constellation from one upward in order of their increasing right ascension. Flamsteed's numbers were used in several catalogues throughout the 18th century, and were expanded to cover all of the sky.

Since Flamsteed's time, astronomers have published countless other star catalogues, and introduced new conventions for naming different kinds of objects – such as Messier (M) and New General Catalogue (NGC) numbers for bright star clusters, nebulae and galaxies, and capital letters for variable stars. When a bright object is catalogued several times, the first name given to it is typically used. So the 24 brightest stars in a constellation are known by their Bayer letters, and the next brightest by their Flamsteed numbers. Some bright star clusters were originally catalogued as stars by Bayer or his followers, so they are still known by stellar names, such as Omega Centauri and 47 Tucanae.

Constellation maps The maps in this book show all the stars above magnitude 6.0 – those within naked-eye visibility on a dark, clear night. To help find your bearings, they are overlaid with a grid showing every 15 degrees of declination and every hour of right ascension. For simplicity, the constellation boundaries are not shown.

THE GREEK ALPHABET

α	alpha	η	eta	ν	nu	τ	tau
β	beta	θ	theta	ξ	xi	υ	upsilon
γ	gamma	ι	iota	ο	omicron	φ	phi
δ	delta	κ	kappa	π	pi	χ	chi
ε	epsilon	λ	lambda	ρ	rho	ψ	psi
ζ	zeta	μ	mu	σ	sigma	ω	omega

Star letters Stars are often designated with Greek letters rather than names. On the pages that follow, these letters are used extensively. This table shows the Greek alphabet with its classical letters and their names.

See also: The changing sky 14 · The celestial sphere 16 · Navigating the skies 18 · The starry sky 82

Signposts of the southern sky

Finding your way around the equator and southern hemisphere *The easily recognized constellation Orion, the Hunter, is the key to many of the stars gathered around the celestial equator on January and February evenings. The most useful markers for far southern skies are Crux, the Southern Cross, and the misleading False Cross pattern.*

Orion group Orion's right shoulder is marked by the red star Betelgeuse (Alpha Orionis), and his left knee by the blue star Rigel (Beta Orionis). A line through the three stars of the belt points north to the red giant Aldebaran (Alpha Tauri) and south to Sirius (Alpha Canis Majoris), the brightest star in the sky. A line across his shoulders points west to Procyon (Alpha Canis Minoris), and east toward Cetus, while a vertical line from Kappa Orionis through Sigma, the westernmost star of the belt, points north to Capella (Alpha Aurigae). Northeast of Orion lies Gemini, with its twin stars Castor and Pollux (Alpha and Beta Geminorum).

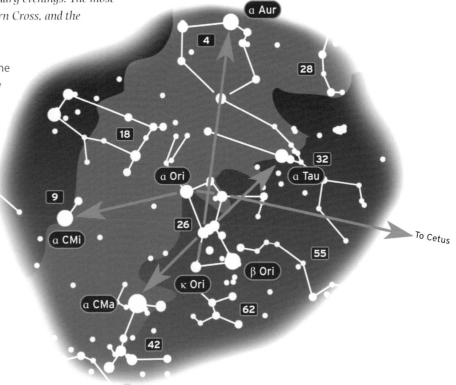

4 AURIGA	**32** TAURUS
9 CANIS MINOR	**42** CANIS MAJOR
18 GEMINI	**55** ERIDANUS
26 ORION	**62** LEPUS
28 PERSEUS	

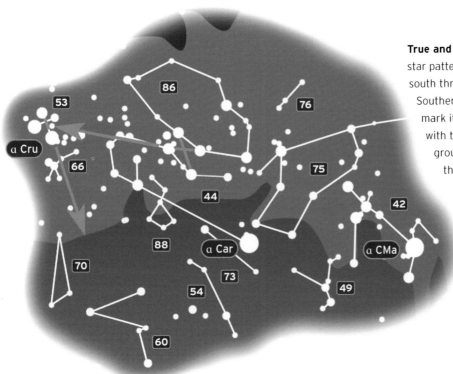

True and false crosses The southern sky's most distinctive star pattern is the Southern Cross, Crux. A line drawn south through Acrux, its brightest star, points toward the Southern Celestial Pole, though there is no bright star to mark its position even roughly. Crux is often confused with the False Cross, a slightly fainter and looser grouping of stars in a similar pattern to the east, on the border of Carina and Vela.

42 CANIS MAJOR	**70** OCTANS
44 CARINA	**73** PICTOR
49 COLUMBA	**75** PUPPIS
53 CRUX	**76** PYXIS
54 DORADO	**86** VELA
60 HYDRUS	**88** VOLANS
66 MICROSCOPIUM	

22 Northern winter and spring

In winter

On long winter nights, northern hemisphere skies are dominated by a fine view of the bright constellation Orion and its retinue. The Milky Way runs across the sky from north to south, while Ursa Major climbs up from the northeastern horizon.

December to February

1 am, December 1

11 pm, January 1

9 pm, February 1

Star Maps for Latitudes 40°-60°N

1	**ANDROMEDA**
5	**BOÖTES**
6	**CAMELOPARDALIS**
8	**CANES VENATICI**
10	**CASSIOPEIA**
11	**CEPHEUS**
14	**CYGNUS**
16	**DRACO**
21	**LEO**
23	**LYNX**
27	**PEGASUS**
28	**PERSEUS**
34	**URSA MAJOR**
35	**URSA MINOR**

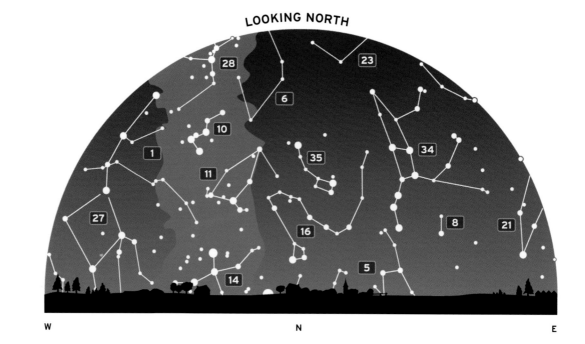

3	**ARIES**
4	**AURIGA**
7	**CANCER**
9	**CANIS MINOR**
18	**GEMINI**
21	**LEO**
26	**ORION**
28	**PERSEUS**
29	**PISCES**
32	**TAURUS**
33	**TRIANGULUM**
42	**CANIS MAJOR**
46	**CETUS**
49	**COLUMBA**
55	**ERIDANUS**
59	**HYDRA**
62	**LEPUS**
75	**PUPPIS**

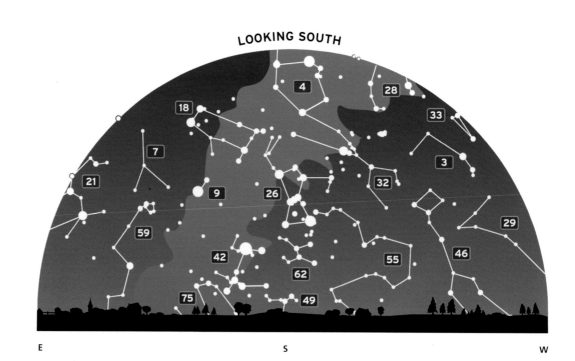

See also: Northern summer and autumn 24 · Southern summer and autumn 26 · Southern winter and spring 28

In spring

On northern spring evenings, Ursa Major lies high overhead, while the view to the south is dominated by the zodiac constellations Cancer, Leo and Virgo. The Milky Way is not prominent before midnight as it runs close to the horizon.

March to May

1 am, March 1
11 pm, April 1
9 pm, May 1

Star Maps for Latitudes 40°–60°N

4 AURIGA
10 CASSIOPEIA
11 CEPHEUS
14 CYGNUS
16 DRACO
18 GEMINI
19 HERCULES
23 LYNX
24 LYRA
25 OPHIUCHUS
26 ORION
28 PERSEUS
32 TAURUS
34 URSA MAJOR
35 URSA MINOR

LOOKING NORTH

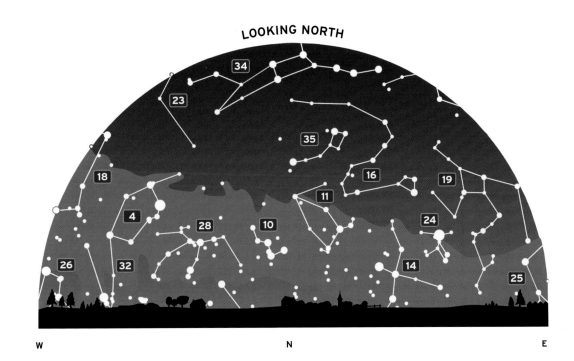

LOOKING SOUTH

5 BOÖTES
7 CANCER
8 CANES VENATICI
9 CANIS MINOR
12 COMA BERENICES
13 CORONA BOREALIS
18 GEMINI
21 LEO
25 OPHIUCHUS
31 SERPENS
34 URSA MAJOR
51 CORVUS
59 HYDRA
63 LIBRA
87 VIRGO

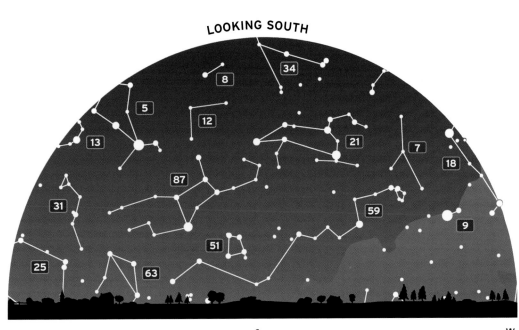

Northern summer and autumn

In summer

Northern summer skies are dominated by the rich star fields of the Milky Way. The finest areas are bounded by the three bright stars of the Summer Triangle, while the constellations of Scorpius and Sagittarius are at their best visibility on the southern horizon.

June to August

1 am, June 1

11 pm, July 1

9 pm, August 1

Star Maps for Latitudes 40°–60°N

1	ANDROMEDA
4	AURIGA
8	CANES VENATICI
10	CASSIOPEIA
11	CEPHEUS
16	DRACO
21	LEO
23	LYNX
27	PEGASUS
28	PERSEUS
33	TRIANGULUM
34	URSA MAJOR
35	URSA MINOR

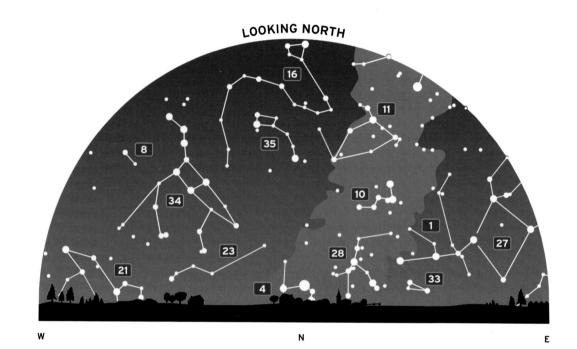

LOOKING NORTH

W   N   E

2	AQUILA
5	BOÖTES
13	CORONA BOREALIS
14	CYGNUS
15	DELPHINUS
19	HERCULES
24	LYRA
25	OPHIUCHUS
27	PEGASUS
30	SAGITTA
31	SERPENS
39	AQUARIUS
43	CAPRICORNUS
63	LIBRA
78	SAGITTARIUS
79	SCORPIUS
87	VIRGO

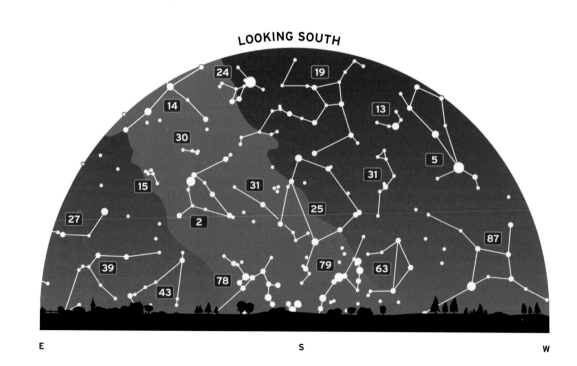

LOOKING SOUTH

E   S   W

See also: Northern winter and spring 22 · Southern summer and autumn 26 · Southern winter and spring 28

In autumn
Autumn skies in the northern hemisphere are relatively barren, with the southern aspect dominated by the large, empty expanse of the square of Pegasus. However, the Milky Way still stretches from east to west overhead and Cassiopeia is at its highest.

September to November
1 am, September 1
11 pm, October 1
9 pm, November 1

Star Maps for Latitudes 40°–60°N

- **4** AURIGA
- **5** BOÖTES
- **10** CASSIOPEIA
- **11** CEPHEUS
- **13** CORONA BOREALIS
- **14** CYGNUS
- **16** DRACO
- **18** GEMINI
- **19** HERCULES
- **23** LYNX
- **24** LYRA
- **25** OPHIUCHUS
- **26** ORION
- **28** PERSEUS
- **34** URSA MAJOR
- **35** URSA MINOR

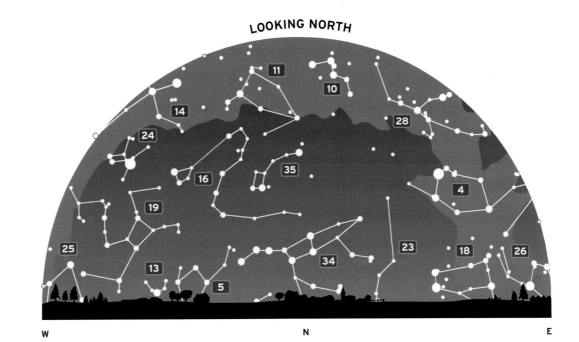

LOOKING NORTH

W   N   E

- **1** ANDROMEDA
- **2** AQUILA
- **3** ARIES
- **14** CYGNUS
- **15** DELPHINUS
- **26** ORION
- **27** PEGASUS
- **29** PISCES
- **30** SAGITTA
- **31** SERPENS
- **32** TAURUS
- **33** TRIANGULUM
- **39** AQUARIUS
- **43** CAPRICORNUS
- **46** CETUS
- **55** ERIDANUS
- **74** PISCIS AUSTRINUS
- **80** SCULPTOR

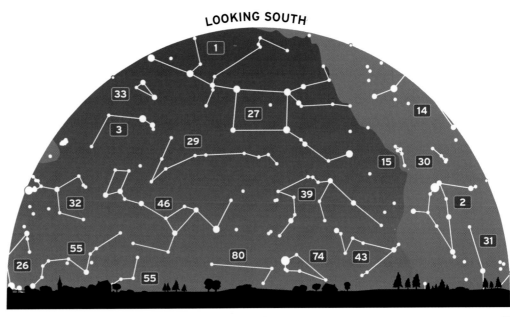

LOOKING SOUTH

E   S   W

Southern summer and autumn

In summer

Southern summer skies offer fine views of the Orion group to the north, with Sirius and Canopus - the sky's two brightest stars - both near the zenith. The southern outlook is dominated by the Milky Way, with the rich constellations of the Argo group and Crux.

December to February

1 am, December 1
11 pm, January 1
9 pm, February 1

Star Maps for Latitudes 20°–40°S

- **3** ARIES
- **4** AURIGA
- **7** CANCER
- **9** CANIS MINOR
- **18** GEMINI
- **21** LEO
- **23** LYNX
- **26** ORION
- **28** PERSEUS
- **29** PISCES
- **32** TAURUS
- **33** TRIANGULUM
- **42** CANIS MAJOR
- **46** CETUS
- **55** ERIDANUS
- **59** HYDRA
- **62** LEPUS

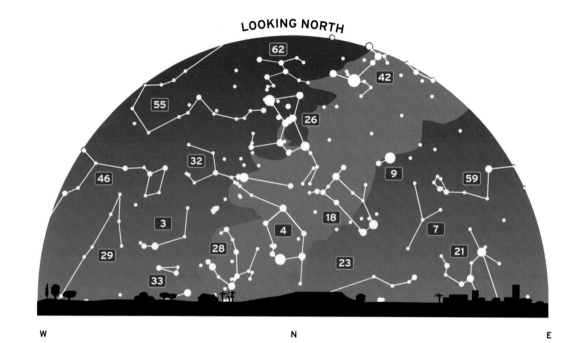

- **44** CARINA
- **45** CENTAURUS
- **49** COLUMBA
- **51** CORVUS
- **53** CRUX
- **54** DORADO
- **55** ERIDANUS
- **57** GRUS
- **59** HYDRA
- **60** HYDRUS
- **61** INDUS
- **66** MICROSCOPIUM
- **70** OCTANS
- **71** PAVO
- **72** PHOENIX
- **73** PICTOR
- **74** PISCIS AUSTRINUS
- **75** PUPPIS
- **76** PYXIS
- **80** SCULPTOR

- **84** TRIANGULUM AUSTRALE
- **85** TUCANA
- **86** VELA
- **88** VOLANS

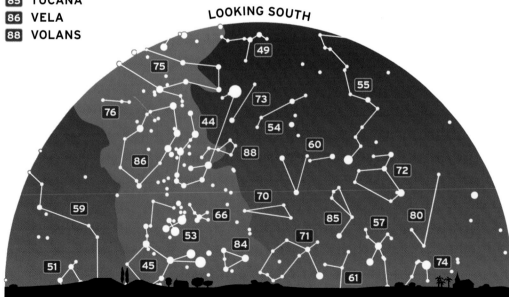

See also: Northern winter and spring 22 · Northern summer and autumn 24 · Southern winter and spring 28

In autumn

Northern skies are relatively barren at this time of year, though Leo, Cancer and Boötes offer some fine sights. To the south, the southern Milky Way is at its finest, with Scorpius, Centaurus, Vela, Carina and Puppis stretching across the sky from east to west.

March to May

1 am, March 1

11 pm, April 1

9 pm, May 1

Star Maps for Latitudes 20°-40°S

5 BOÖTES
7 CANCER
8 CANES VENATICI
9 CANIS MINOR
12 COMA BERENICES
13 CORONA BOREALIS
18 GEMINI
21 LEO
23 LYNX
25 OPHIUCHUS
31 SERPENS
34 URSA MAJOR
51 CORVUS
59 HYDRA
63 LIBRA
87 VIRGO

LOOKING NORTH

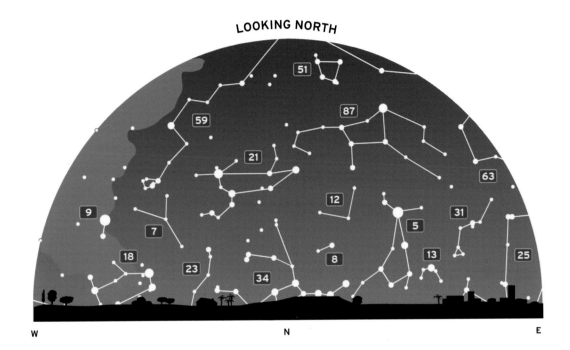

25 OPHIUCHUS
26 ORION
31 SERPENS
40 ARA
42 CANIS MAJOR
44 CARINA
45 CENTAURUS
49 COLUMBA
53 CRUX
54 DORADO
59 HYDRA
60 HYDRUS
62 LEPUS
63 LIBRA
64 LUPUS
66 MICROSCOPIUM
70 OCTANS
71 PAVO
73 PICTOR
75 PUPPIS

76 PYXIS
78 SAGITTARIUS
79 SCORPIUS
84 TRIANGULUM AUSTRALE
85 TUCANA
86 VELA
88 VOLANS

LOOKING SOUTH

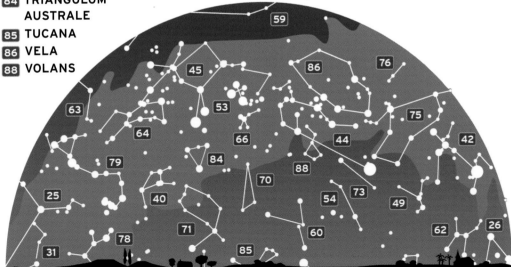

28 Southern winter and spring

In winter

To the north, constellations such as Cygnus, Lyra and Hercules reach their best visibility at this time of year, while the Milky Way stretches across the sky from north to south, with the rich star clouds of Sagittarius and Scorpius directly overhead.

June to August

1 am, June 1

11 pm, July 1

9 pm, August 1

Star Maps for Latitudes 20°-40°S

- **2** AQUILA
- **5** BOÖTES
- **13** CORONA BOREALIS
- **14** CYGNUS
- **15** DELPHINUS
- **16** DRACO
- **19** HERCULES
- **24** LYRA
- **25** OPHIUCHUS
- **27** PEGASUS
- **29** PISCES
- **30** SAGITTA
- **31** SERPENS
- **39** AQUARIUS
- **63** LIBRA
- **78** SAGITTARIUS
- **79** SCORPIUS
- **87** VIRGO

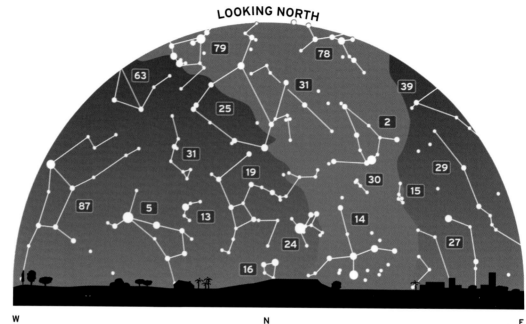

LOOKING NORTH

W   N   E

- **39** AQUARIUS
- **40** ARA
- **43** CAPRICORNUS
- **44** CARINA
- **45** CENTAURUS
- **51** CORVUS
- **53** CRUX
- **55** ERIDANUS
- **57** GRUS
- **59** HYDRA
- **60** HYDRUS
- **61** INDUS
- **64** LUPUS
- **66** MICROSCOPIUM
- **70** OCTANS
- **71** PAVO
- **72** PHOENIX
- **74** PISCIS AUSTRINUS
- **75** PUPPIS
- **78** SAGITTARIUS

- **79** SCORPIUS
- **80** SCULPTOR
- **84** TRIANGULUM AUSTRALE
- **85** TUCANA
- **88** VOLANS

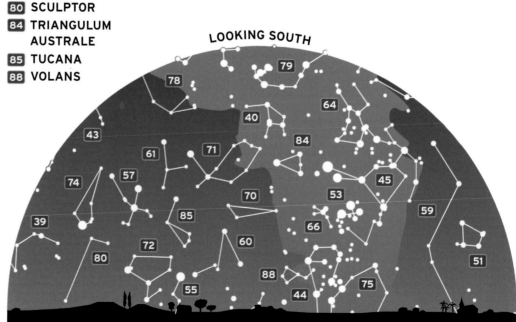

LOOKING SOUTH

E   S   W

See also: Northern winter and spring 22 · Northern summer and autumn 24 · Southern summer and autumn 26

In spring

In southern spring, Pegasus reaches its highest point to the north, bringing with it fairly empty skies, but fine views of galaxies beyond our own. To the south, the bright constellations are at their lowest and the sky is dominated by the faint Southern Birds.

September to November

1 am, September 1

11 pm, October 1

9 pm, November 1

Star Maps for Latitudes 20º-40ºS

1. ANDROMEDA
2. AQUILA
3. ARIES
14. CYGNUS
15. DELPHINUS
26. ORION
27. PEGASUS
28. PERSEUS
29. PISCES
30. SAGITTA
31. SERPENS
32. TAURUS
33. TRIANGULUM
39. AQUARIUS
43. CAPRICORNUS
46. CETUS
55. ERIDANUS
74. PISCIS AUSTRINUS

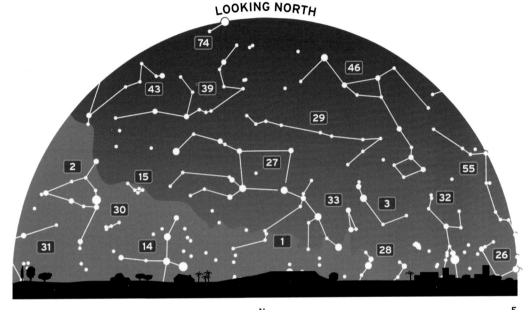

LOOKING NORTH

W   N   E

25. OPHIUCHUS
26. ORION
40. ARA
43. CAPRICORNUS
44. CARINA
45. CENTAURUS
49. COLUMBA
54. DORADO
55. ERIDANUS
57. GRUS
60. HYDRUS
61. INDUS
62. LEPUS
64. LUPUS
66. MICROSCOPIUM
70. OCTANS
71. PAVO
72. PHOENIX
73. PICTOR
74. PISCIS AUSTRINUS

75. PUPPIS
78. SAGITTARIUS
79. SCORPIUS
80. SCULPTOR
84. TRIANGULUM AUSTRALE
85. TUCANA
88. VOLANS

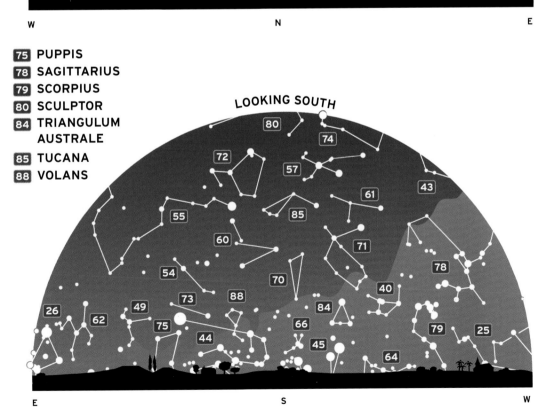

LOOKING SOUTH

E   S   W

Saturnrise Seen from above the cloud-tops of its giant moon Titan, Saturn hangs perpetually on the horizon. At dawn its bloated crescent is bisected by the narrow line of its rings as Titan passes through their plane. Although Titan appears to be an idyllic world, perhaps suitable for life, it is in fact bitterly cold, with clouds made up of poisonous and highly flammable methane – only stable because the moon is completely lacking in oxygen.

The Solar System

32 The home system

The kingdom of the Sun is a tiny backwater on the cosmic scale, though vast in comparison with the familiar distances of Earth. Our home system extends to an immense sphere stretching ten trillion kilometres beyond the orbit of distant Pluto.

The largest objects in orbit around the Sun are the planets. These are ordered into groups with four small, rocky worlds close to the Sun, four giant balls of gas much farther away, plus Pluto – a tiny, enigmatic ball of ice that spends most of its time in the depths of space, farthest from the Sun.

Most of the planets have moons, which range in size from tiny lumps of rock to complex worlds larger than Pluto and Mercury. In addition, the four gas-giant planets have ring systems, made up of billions of small particles.

In between and beyond the planets lie vast collections of smaller bodies. Close to the Sun, these are lumps of rock called asteroids, and farther out they are chunks of deep-frozen ice— dormant comets that become active only when their orbits bring them near the Sun's warmth. Most of these small bodies are confined to three major zones: an asteroid belt between the orbits of Mars and Jupiter, the Kuiper Belt beyond Neptune – which includes Pluto – and the Oort Cloud, a vast hollow shell of deep-frozen comets far out in space at the limits of the Sun's gravitational influence.

Where is the edge of the Solar System? Some astronomers place it at the heliopause. This is a region where the solar wind – the stream of particles blown out into space by the Sun – falters and dies in the face of the pressure of a billion other stellar winds blowing across the galaxy. However, the influence of the Sun stretches far beyond the heliopause: its gravitational attraction is the most significant force throughout a shell of space roughly two light-years across.

Leaving home Looking back across the Solar System, we see that all the planets except Pluto orbit close to a flat plane stretching out from the Sun's equator.

See also: Genesis 34 · Other solar systems 100 · The Copernican revolution 192 · Probes to the planets 202

The zodiac

The belt of living things *Twelve of the oldest constellations lie in a belt that circles the sky, called the zodiac. They mark the course of the ecliptic – the path that the Sun takes around the sky in one year. This is equivalent to the plane of Earth's orbit around the Sun, and all the planets, except for Pluto, have orbits that lie close to it. As a result of this, the Sun, and most planets can* *always be found somewhere along it – a fact that gave the zodiac special significance in the eyes of early astronomers. The Moon also orbits close to the ecliptic and so is often found near the zodiac. Every year, the Sun moves through all 12 constellations of the zodiac, but it also passes through part of a 13th figure—Ophiuchus, the Serpent-Bearer.*

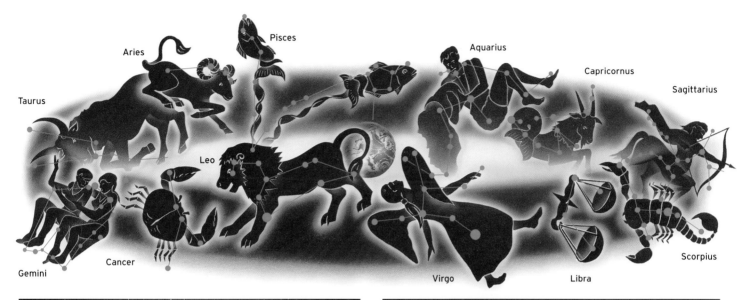

Zodiacal light On very dark, moonless nights, it may be possible to spot a dim triangle of soft light rising above the western horizon after sunset, or in the east before sunrise. This glow lies along the ecliptic and (in the northern hemisphere) is best seen in spring after sunset or before sunrise in autumn. It comes from sunlight reflecting off dust particles (left by comets or asteroid collisions) in the plane of the Solar System.

Planetary conjunctions Because the planets orbit in roughly the same plane, they can come close to each other in the sky - an event called a conjunction. Here, in close conjunction from left to right, are the Moon, Mars, Jupiter, and Saturn. The Moon also orbits close to the ecliptic and is often involved in conjunctions. Sometimes it can even pass in front of a planet - an event called an occultation.

34 Genesis

The Solar System was born out of fire and ice, in orbit around a young star about 4.6 billion years ago. Astronomers have dated its formation and chronicled its early history by studying radioactive elements in the rocks of Earth, the Moon and meteorites.

The Sun and its family were created when the solar nebula – a cloud of interstellar gas and dust – began to collapse about 5 billion years ago. This event was probably triggered by the violent death of a nearby star in an enormous supernova explosion. The solar nebula was almost entirely made of hydrogen and helium. There was also a tiny but vital amount of heavier elements such as oxygen, carbon, nitrogen, and metals created in the nuclear furnaces of an earlier generation of stars.

As the nebula collapsed, it began to spin more rapidly. At the same time, the cloud began to flatten into a disc, encircling an increasingly hot ball of gas that would eventually become our Sun.

But even before the infant Sun began to shine as a star, something was stirring out in the solar nebula. Relatively heavy particles of dust and metal that constantly collided with each other occasionally stuck together. Small clumps of dust slowly grew larger, eventually forming planetesimals – lumps of rock with enough gravity to pull themselves into a rough sphere and draw in more material from their surroundings. In time, these planetesimals swept space clear of most of the dust. Occasionally they would collide with each other as well. The heat released by these collisions, and the constant rain of new material, slowly heated the young planets to melting point, allowing dense metals such as iron to sink to their cores.

As the Sun began fusing hydrogen in its core, its radiation swept away most of the light gases from the inner Solar System, leaving just the rocky planets. Farther out, where temperatures were much colder, gases remained abundant. Here, the growth of planets involved large quantities of gas along with ice and rocky materials. Eventually, like solar systems in miniature, these condensed into the gas-giant planets and their families of moons.

Although the eight major worlds of the Solar System formed rapidly, there were still large amounts of debris floating between them. Over hundreds of millions of years, much of this debris collided with planets or moons. Rocky fragments remain orbiting in the asteroid belt between Mars and Jupiter, while icy bodies fill the Kuiper Belt beyond Neptune. Other pieces of debris have been thrown out of the Solar System, and some went to populate the far-flung Oort cloud of comets that extends as far as the Sun's gravity exerts its influence. Most of these small bodies, rocky or icy, are primitive remnants of the Solar System's birth.

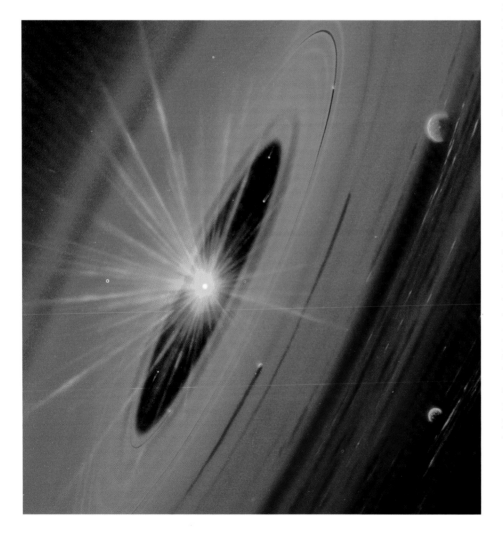

A star is born The protostar – our young Sun – begins to shine as planetesimals grow by sweeping the space around them clean.

See also: The home system 32 · Star birth 94 · Other solar systems 100 · Clouds in space 110

The debris of creation

Postcards from the past *Although most of the debris from the birth of the Solar System was swept up billions of years ago, there is still plenty of it out there – in the form of comets, asteroids and meteorites. A meteorite is a lump of space rock large enough to survive a journey through Earth's atmosphere and make it to the surface intact. They offer astronomers a unique glimpse of the material that formed the Solar System. Because they have changed only minimally, most meteorites are samples of the nebula from which the Sun and the planets were born.*

Chondrite meteorites consist of thousands of tiny spheres, or chondrules. These meteorites are believed to be the most primitive, dating back to the Solar System's birth – the chondrules formed as molten droplets, condensing like raindrops from the solar nebula.

Stony-iron meteorites contain large crystals of metals such as iron and nickel, embedded in a stony matrix. These metals sink to the centre of any body large enough to melt completely, so these meteorites are believed to originate around the cores of primitive planetesimals.

Fearsome visitors Falling meteorites often create fireballs - shooting stars much brighter than those created by smaller dust particles. The most spectacular fireballs of all are bolides, brilliant, slow-moving objects, which are often accompanied by a sonic boom.

The meteorite hunters Unless someone actually sees it fall, a meteorite can often be very hard to identify. Meteorite hunters have learned to concentrate on looking in regions such as the Sahara Desert and Antarctica, where surface rocks are easy to spot. Rocks found in these regions have a higher probability of being extraterrestrial. Here, a NASA robot called Nomad locates a meteorite on an Antarctic ice field.

Nickel-iron meteorites are the remnants of planetesimal cores - large lumps of metal with a mineral crust. They contain large metal crystals, indicating that they cooled very slowly.

Our star, the Sun

36

The Sun is a hot ball of hydrogen gas 1.4 million km (800,000 miles) across, and 150 million km (93 million miles) from Earth. The immense temperatures and pressures in its core cause nuclear fusion reactions that make it shine.

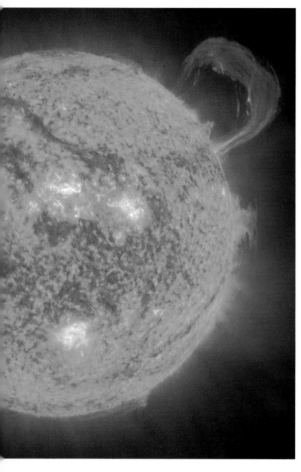

Fiery orb Seen at ultraviolet wavelengths, the Sun's photosphere seethes with activity as radiation blasts out from within.

Seen from a distance, the Sun is a ball of blinding yellow-white light, with an apparently perfect surface. But photographs of the nonvisible radiation from the Sun reveal that it is far from calm and constant. Pulsating, dark-edged granules and pillars of flame many thousands of miles high betray the turbulence below the surface.

The Sun's interior is trapped in a delicate equilibrium, poised between collapsing inward under its own weight and blowing itself apart from the pressure at the core, and the sheer force of the radiation blasting out through it. Temperatures at the core reach 15 million °C (27 million °F), triggering nuclear reactions that release gamma rays – high-energy radiation that would be deadly if it poured out of the Sun unaltered. Fortunately for life on Earth, the core occupies only a quarter of the Sun's diameter, and it is surrounded by other layers that absorb much of this lethal energy.

Temperatures are so high inside the Sun that hydrogen atoms split apart forming an electrically charged plasma of atomic nuclei and electrons. Within the radiative zone – the layer surrounding the core – the plasma is so highly compressed that gamma rays from the core can only travel for tiny distances before colliding with something. Each collision absorbs some of the rays' original energy, keeping the radiative zone hot, and supporting the star. Radiation takes up to 100,000 years to escape the radiative zone, slowly losing its energy as it goes.

Then, 130,000 km (80,000 miles) below the Sun's surface, the escaping radiation hits a barrier, as the temperature drops low enough for hydrogen atoms to be stable. This hydrogen envelope – the convective zone – is opaque, and soaks up the radiation as it pours out from the interior. As large volumes of gas heat up, they start to rise through their cooler surroundings by convection, at temperatures of around 5,500 °C (9,900 °F). Farther out, the density drops, until suddenly the Sun becomes transparent again. Radiation bursts out of the gas, mostly as visible light, ultraviolet and infrared radiation.

The photosphere, where the Sun becomes transparent, forms the visible surface of the Sun, but it is not its edge. Directly above the photosphere, a vast but tenuous atmosphere surrounds the Sun – the chromosphere and the corona, where a huge halo of gas heated to millions of degrees streams out, eventually merging with the solar wind.

SUN PROFILE

Equatorial diameter	1.392 million km		Mass	2×10^{27} tonnes	
	864,900 miles		Mass (Earth = 1)	333,000	
Surface temperature	5,500 °C		Power output	3.8×10^{26} W	
	9,900 °F		Life expectancy	5 billion Earth years	
Core temperature	15 million °C		Rotation period (polar)	34 Earth days	
	27 million °F		Rotation period (equatorial)	25 Earth days	

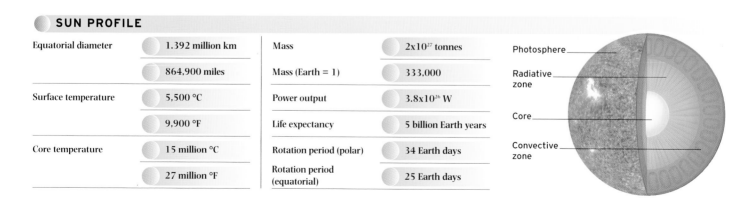

Photosphere
Radiative zone
Core
Convective zone

See also: The solar cycle 38 · Nearby stars 86 · Star birth 94 · Young stars 96 · Sunlike stars 98

Observing the Sun

The Sun and the seasons *The Sun's track across the sky changes day by day—the tilt of Earth's axis means it moves from 23.5 degrees south to 23.5 degrees north of the celestial equator and back once a year. This is most obvious at middle and high latitudes. At latitude 50 degrees, for example, the Sun reaches 63.5 degrees above the horizon at noon in midsummer, but only 16.5 degrees in midwinter.*

The Sun is high in the sky in the northern hemisphere summer.

Sun

Short shadow

Shadow on ground

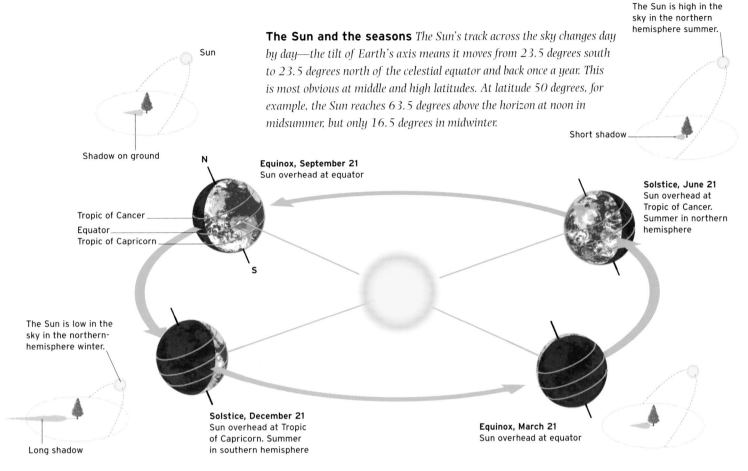

N

Equinox, September 21
Sun overhead at equator

Tropic of Cancer

Equator

Tropic of Capricorn

S

Solstice, June 21
Sun overhead at
Tropic of Cancer.
Summer in northern
hemisphere

The Sun is low in the sky in the northern-hemisphere winter.

Solstice, December 21
Sun overhead at Tropic
of Capricorn. Summer
in southern hemisphere

Equinox, March 21
Sun overhead at equator

Long shadow

Watch out for your eyes The Sun is always far too bright to look at directly – even at sunset, staring at it can permanently damage your eyes. Never look at the Sun through binoculars or a telescope that is not equipped with a proper aluminized solar filter that fits over the front of the instrument. This reduces the Sun's light before it enters the optics. Do not use a filter that fits over the eyepiece - these can crack without warning and blind you. Another safe way to view the Sun is by projection. Put a cheap eyepiece into the focuser, aim the telescope at the Sun by looking at its shadow, and place a white card behind the eyepiece, adjusting the focus as needed. However, do this only with a refractor or Newtonian reflector telescope of 100 mm (4 in) aperture or less. You can mask a larger aperture to reduce its size. Avoid projection altogether with binoculars and Schmidt-Cassegrain or Maksutov telescopes as the Sun will heat their optics too much.

STAR STORIES

Stonehenge
The rising and setting points of the Sun move back and forth along the horizon during the year as the Sun moves higher or lower in the sky. Many ancient monuments around the world were apparently designed with alignments to these points at important times of year such as the equinoxes or solstices. At Stonehenge in England the midsummer Sun rises over the outlying "Heel Stone", whereas at the great temple of Karnak, in Egypt, the midwinter sunrise was visible from a chapel called the "High Room of the Sun".

The solar cycle

About every 11 years, the Sun reaches peak activity – sunspots grow in number and size, huge prominences arc into space, and solar flares can erupt with enough power to damage satellites in orbit around Earth.

Ultra-violence on the surface of the Sun Loops of hot, dense gas bursting into the corona reveal the loops of the Sun's magnetic field in this ultraviolet photograph.

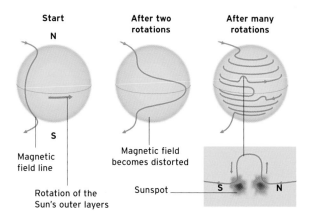

Start

N

S

Magnetic field line

Rotation of the Sun's outer layers

After two rotations

Magnetic field becomes distorted

Sunspot

After many rotations

S   N

The tortured magnetic field of the Sun The Sun's magnetic field is twisted and wound up because the outer layers of the Sun rotate faster near the equator than near the poles. Successive rotations wrap the field around the Sun, pushing loops up through the photosphere to form sunspots and prominences.

Sunspots, prominences – huge loops of hot gas arcing from the Sun's surface – and solar flares are all linked to changes in the Sun's magnetic field. The Sun's magnetism is generated by movement of electrically charged gas, or plasma, within it. Unlike Earth, the Sun is completely fluid, so different parts of it can move at different rates. Regions near the poles rotate about once every 35 days, whereas regions close to the equator rotate once every 25 days.

This differential rotation seems to occur only in the convective zone – the Sun's outer layer. Deeper inside, the Sun behaves like a solid body. As the convective zone rotates, it carries the magnetic field with it, twisting it into a tangled pattern. This becomes concentrated in certain regions and parts of it start to cancel each other out. The Sun's magnetism briefly fades to nothing before it regenerates. This solar cycle creates and destroys the solar magnetic field about every 11 years and flips over the Sun's magnetic poles.

The most violent solar activity occurs where the magnetic field loops up into space. Where the ends of the loop pass through the photosphere, they create cooler regions of lower density – sunspots. Sunspots appear dark compared with the hotter and brighter regions around them, but they are still very hot, around 3,500 °C (6,300 °F).

Above the sunspots, gas streams outward along the loops in the magnetic field, forming prominences. The gas in a prominence is cooler but denser and therefore brighter than its surroundings. Solar flares occur when magnetic loops meet – they short-circuit, releasing huge amounts of energy. This is constantly happening on a small scale, but large-scale reconnections can create vast flares or coronal mass ejections that send billions of tons of hot gas hurtling into space.

See also: Our star, the Sun 36 · The wind from the Sun 42 · Atmospheric phenomena 46 · Sunlike stars 98

Following solar activity

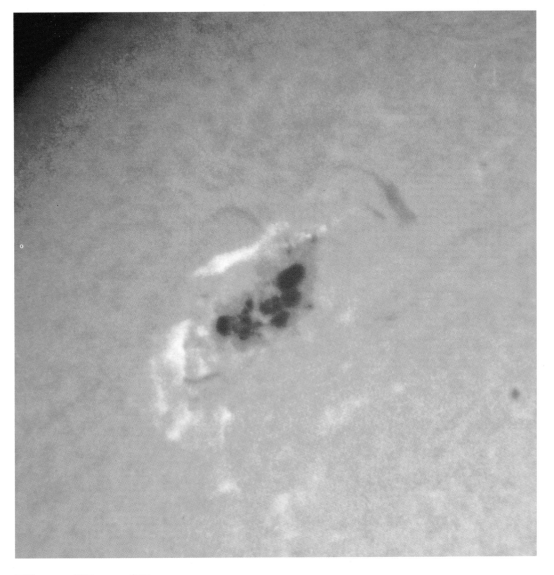

Colours indicate the percentage of the Sun's surface covered with sunspots.

>0.0%   >0.1%   >1.0%

Year

1900   1920   1940   1960   1980   2000

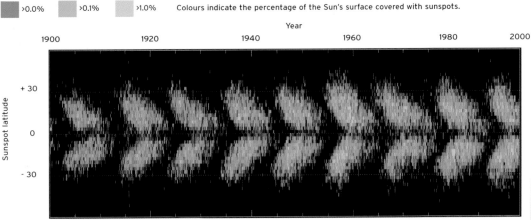

Sunspot latitude

+ 30

0

- 30

A century of sunspots (1900-2000) By recording the amount of sunspots and their positions regularly over several years, it is easy to see a clear pattern emerging. It was this diagram, with its neat sets of butterfly patterns, that led to the discovery of the solar cycle. Each cycle begins with the appearance of small numbers of sunspots at high latitudes. As the cycle continues, sunspots appear and vanish at ever lower latitudes. Eventually, as the field begins to fade, sunspot numbers dwindle away - yet, the first spots of the next cycle are already appearing at high latitudes.

Blemishes on the face of the Sun *Sunspots always occur in pairs – a preceding spot that moves ahead and is slightly closer to the equator, and a trailing spot that follows behind and is closer to the pole. The two spots in the pair have opposite magnetic polarities and are linked by an invisible loop of magnetic field above them.*

This image of a sunspot group shows the structure clearly. Each spot has a dark central shadow, or umbra, where it is coolest, and a paler surrounding shadow, the penumbra. Close up, the edges of the penumbra look rather like iron filings placed close to a magnet.

✦ STAR STORIES

The Little Ice Age

Although a link between the solar cycle and Earth's weather has yet to be proved, there is good evidence that the cycle has affected our climate in the past. The first sunspots were recorded around 1610 but, between 1645 and 1715, almost no spots were seen. This period, known as the Maunder Minimum, after the astronomer who first noticed it, coincided with the Little Ice Age of exceptionally cold winters in Europe and elsewhere. In London, the Thames River froze regularly, and frost fairs were held on the ice.

40 Solar eclipses

Solar eclipses are one of nature's most spectacular special effects. As the Sun is blotted out and night falls across the landscape in seconds, it is easy to understand the fear these events aroused in our ancestors.

Solar eclipses happen when the Sun and Moon line up precisely in an Earthbound observer's line of sight. Although the Moon is 400 times smaller than the Sun, it is also about 400 times closer to Earth. This enables a total eclipse to occur, while allowing us to see details that are very close to the Sun's edge.

Eclipses are rare events because the Moon's orbit is tilted to the Sun's apparent path across the sky. In any one month, the Moon only crosses the ecliptic at two points, and there is no guarantee that the Sun will be close to these points at the same time as the Moon.

Furthermore, because the Moon is so small, it casts a very narrow shadow on Earth and so total solar eclipses occur over very small sections of Earth's surface. Seen from space, the eclipse casts a dark central shadow – the umbra – onto Earth, surrounded by a ring of fainter shadow – the penumbra. This shadow moves across the landscape at a speed of 1,600 km/h (1,000 mph) or more, tracing out an eclipse track hundreds of kilometres long, but usually just a few tens of kilometres wide.

The length of time that the Sun is completely blocked out depends on the observer's position on the eclipse track and also on the changing distance of the Moon from Earth. At its closest to Earth, the Moon can eclipse the Sun for up to seven minutes, but at its most distant, it is not even large enough to cover the Sun completely, and instead creates an annular eclipse where a ring of the Sun shines out from around the silhouetted Moon.

By blocking out the dazzling light from the photosphere, a total eclipse reveals the much fainter outer layers of the Sun's atmosphere. The corona, extending into space for millions of kilometres, consists of sparsely scattered gas at temperatures of 1 to 2 million °C (2 to 4 million °F). Eclipses are a chance to study its shape, and both amateur and professional astronomers relish the opportunity. The shape of the Sun's corona is intrinsically linked to the solar cycle – at times of solar minimum, the corona is small and round, while at solar maximum it is large and uneven.

Because the penumbra is so much larger than the umbra, it covers a bigger area on Earth, and therefore partial eclipses are much more widely seen than total ones. Partial eclipses can even happen when the umbra misses Earth and there is no total eclipse. A partial eclipse also creates some unusual effects, such as the dappled light shining through leaves turning from circular patches to crescent shapes, mimicking the image of the Sun.

Total solar eclipse This composite photograph shows the Moon moving from left to right after a total solar eclipse. As the Sun reappears from behind the Moon, features at the outer edge of the Sun – such as prominences – become visible.

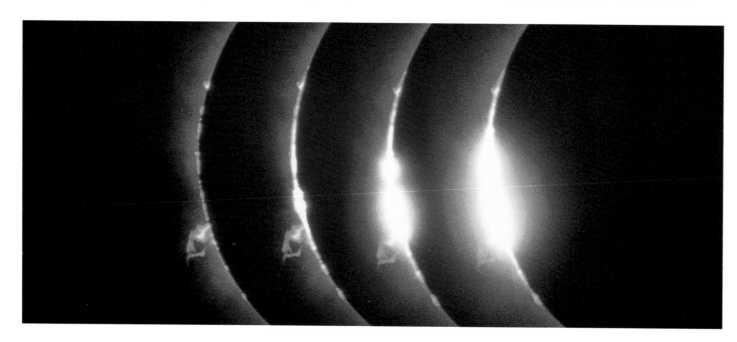

See also: Our star, the Sun 36 · The solar cycle 38 · Amateur astronomy 184

Eclipse watching

Totally amazing *If you're lucky enough to see a total eclipse, there are a number of phenomena both before and during the event that are worth watching out for. As the Sun shrinks to a sliver, strange parallel shadows appear on the ground. These are a result of Earth's unstable atmosphere acting like a lens. The temperature and light levels drop very suddenly in the last couple of minutes before totality, and observers in high places can see the edge of the Moon's shadow rushing toward them. Even when there is cloud cover, the shadow can sometimes be perceived as it passes over the clouds.*

Diamond ring effect The most beautiful moment in a total eclipse happens just before the Moon covers the Sun completely. For a couple of seconds, a final spot on the Sun shines out, creating the awe-inspiring diamond ring effect.

Baily's Beads Sometimes, the last part of the Sun to disappear vanishes behind a range of mountains on the edge of the Moon. When this happens, a series of brilliant points of light shine through the valleys along the Moon's edge for a few seconds. These are known as Baily's Beads.

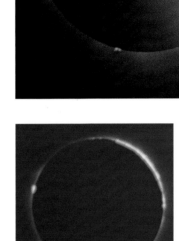

Prominences If an eclipse coincides with a period of high solar activity, the Moon's disc can often be ringed with reddish-coloured prominences – huge loops of hot gas arcing from the surface of the Sun. The corona will also be at its largest when the solar cycle is near to maximum.

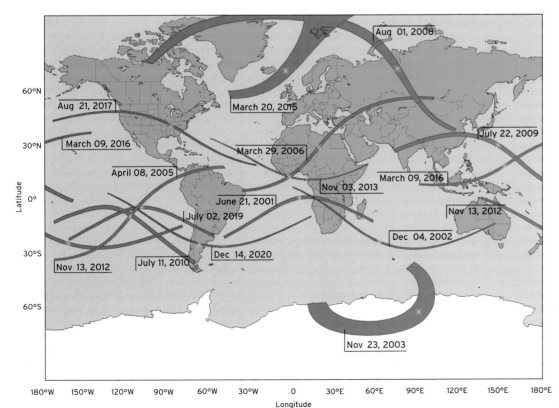

Charting future eclipses On average any place on Earth will see just one eclipse every 300 years. Eclipses recur over a 56-year period, called the Saros. Although this cycle can help to predict the possibility of an eclipse, it cannot guarantee that one will happen, or predict where on Earth one will occur. This map shows the locations of eclipses and their durations between 2001 and 2020.

42 The wind from the Sun

The Sun blasts a hail of high-speed particles across the Solar System from its upper atmosphere. The small amount of this material that makes it through to Earth's atmosphere creates harmless but beautiful displays of light – the aurorae.

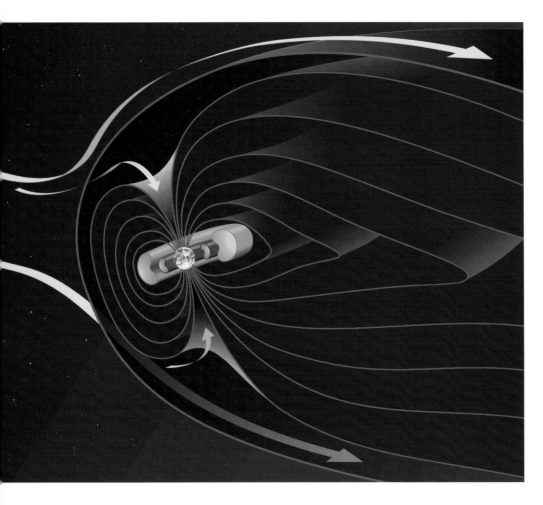

Van Allen belts Most of the solar wind (yellow arrows) is diverted around Earth by the magnetosphere (dark, concentric bubbles), but a small number of particles slowly sift downward, eventually becoming trapped in two doughnut-shaped regions around Earth – the Van Allen radiation belts (red and dark-red doughnuts). Magnetic fields here cause the particles to ricochet back and forth, generating huge electrical currents within the belts.

The solar wind is a mixture of atomic nuclei and subatomic particles blown out from the Sun's corona by the force of radiation. At the edges of the corona, it travels at speeds of up to 900 km/s (1,000 mps), although by the time it reaches Earth, it has slowed to about half this speed.

If these particles could reach the surface of Earth, their effects would be deadly as they slammed into the soft cells of living tissues. Fortunately, though, Earth is protected from the onslaught by its natural magnetic field. The field, generated by currents in the molten core – driven by Earth's spin – surrounds Earth like the field around a bar magnet. Its influence extends thousands of miles into space, where it interacts with the solar wind.

Because the particles in the wind are electrically charged, they are affected by magnetism. As they encounter the outer reaches of Earth's magnetic field, their speed drops rapidly until, at a barrier called the magnetopause, the solar wind comes to a complete halt against the outward pressure of Earth's magnetism.

From the magnetopause, some particles enter the Van Allen belts around Earth. They can only escape from the belts close to Earth's magnetic poles, where they rain down into the upper atmosphere. Here, they collide with gas atoms and molecules, injecting them with additional energy and causing them to glow. The result is an aurora.

Auroral displays are normally seen from locations close to the "auroral ovals" – rings around north and south magnetic poles, normally about 20 degrees away from the magnetic poles themselves. These ovals mark the points on Earth's surface below the charged particles as they pour into the atmosphere. They expand at night when, shielded from the pressure of the solar wind, Earth's magnetic field blossoms outward. Because the magnetic poles are tilted roughly 11 degrees away from the Earth's poles of rotation, the oval dips farthest south over North America.

Individual auroral displays are hard to predict, but they display a long-term cycle that is linked to changes in the Sun itself. About every 11 years, activity on the Sun peaks, and solar flares and coronal mass ejections pour billions of tons of extra particles into the solar wind. Spectacular aurorae are often seen on Earth one or two days after major events.

See also: The home system 32 · Our star, the Sun 36 · The solar cycle 38 · Earth – the dynamic planet 44

THE WIND FROM THE SUN

Observing aurorae

Strange lights at night *Typically, auroral glows first appear in the skies poleward of an observer after nightfall. The glow looks like the beginning of dawn – aurora borealis literally means "northern dawn". During the night, the aurora expands away from the pole, rising up the sky to form an arc – an even arch of light with a smooth lower edge – or a band, with a knotted lower edge. Within this, features often appear, such as flickering patches of light and curtains or vertical rays.*

Auroral coronae A strong auroral display can climb up the poleward sky as far as the zenith, directly overhead. Here, it will form a corona - a burst of rays spreading out in all directions. The shape of Earth's magnetic field is brilliantly illuminated as charged particles stream into the atmosphere from directly above.

Auroral colours Gas atoms in Earth's atmosphere change their structure with altitude, and emit different colours of light. The most common gases, nitrogen and oxygen, emit pinkish light at low altitudes (around 100 km or 60 miles). Nitrogen emits blue and violet light at around 140 km (90 miles), and pink again in the upper atmosphere. Oxygen emits a greenish glow–the most common auroral colour - at around 140 km, and red farther up. Other rarer gases, such as neon, hydrogen and helium, can produce aurorae, but these mostly appear as isolated splashes of light.

44 Earth – the dynamic planet

Our planet is unique in the Solar System. It is the only world able to retain large volumes of liquid water on its surface, and this may well be why life began to evolve here. But this is just one of Earth's many differences from its neighbours.

Earth is a rocky, or terrestrial planet the largest of four in the inner Solar System. Although other worlds of similar size exist farther from the Sun such as the larger moons of the gas-giant planets – they formed in a region of the Solar System where ices could survive without melting and therefore make up a large part of their composition. Earth and its near neighbours, however, formed mostly from rocky material.

As planetesimals come together to form a planet, the collisions generate heat. The bigger a planet, the more collisions it takes to form, the hotter it gets, and the longer it takes to cool down. As a planet heats up and melts, materials are free to circulate, and heavier ones fall toward its centre.

This process, called differentiation, separated Earth's interior into several layers – a central core of molten metal, surrounded by lower and upper mantles of molten and semimolten rock, and a thin crust of solid rock.

Convection keeps the inner regions of Earth in constant motion, and these large-scale movements of the mantle push huge continent-sized plates of crust around, remolding the surface of the planet. Earth is the only planet with such active "tectonics", mainly because it is bigger and has a hotter interior than its smaller neighbours.

Water has also had an enormous effect on the evolution of the planet, and its presence has been crucial in the history of life. Earth developed in a temperate zone of the Solar System, where water could remain liquid on any planet with a reasonably thick protective atmosphere. But any chunks of ice that might have been in this region before Earth formed would have evaporated long before, in the fierce light of the young Sun. So where did Earth's water come from?

One theory is that it arrived later, after formation of the atmosphere, and at about the time the giant-gas planets formed. The chaos in the outer Solar System caused by the formation of these planets, disrupted huge numbers of icy comets orbiting there and sent many of them hurtling toward the Sun. Inevitably some collided with the inner planets, and these might be responsible for the oceans of Earth. The bombardment might also have provided the carbon-based "organic" chemicals required for the creation of life itself.

Earth rising over the Moon Seen from space, Earth is an azure globe adorned with swirling patterns of snow-white clouds. It is an oasis in the barren wastes of the Solar System and, as far as our current knowledge reaches, the only home to life in the Universe.

PLANET PROFILE

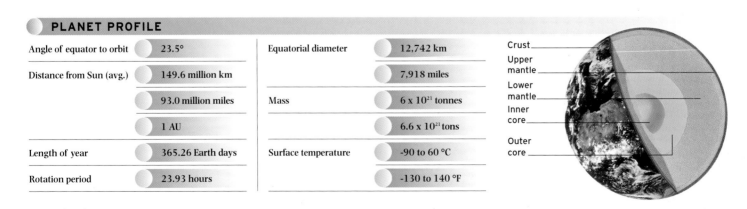

Angle of equator to orbit	23.5°	Equatorial diameter	12,742 km	Crust	
			7,918 miles	Upper mantle	
Distance from Sun (avg.)	149.6 million km			Lower mantle	
	93.0 million miles	Mass	6 x 10²¹ tonnes	Inner core	
	1 AU		6.6 x 10²¹ tons	Outer core	
Length of year	365.26 Earth days	Surface temperature	-90 to 60 °C		
Rotation period	23.93 hours		-130 to 140 °F		

See also: The home system 32 · Genesis 34 · Our star, the Sun 36 · Atmospheric phenomena 46 · The Moon 50

Skylights

Refraction of light *Earth is shrouded in an atmosphere that affects the light passing through it, giving rise to a number of beautiful effects. Earth's atmosphere bends light in the same way as a lens, distorting the Sun's image. Density differences in the thick layer of atmosphere around the horizon can cause the Sun's image to change shape, and even colour, quite rapidly. Different colours of light are refracted by slightly different amounts, so that the bottom edge of the Sun is reddest, whereas the top edge remains yellow longest.*

The setting sun During the day, the Sun is overhead and light from it passes through a relatively thin layer of atmosphere. The blue component of sunlight tends to be scattered and spread across the sky and the Sun appears more yellow than it actually is. Toward sunset, however, the light has to pass through a much thicker layer. Now, not only blue, but also green and yellow light is scattered, creating bands of different colours and turning the Sun red.

Earth's shadow A clear sky at sunset reveals the shape of the Earth. Crepuscular rays - sunlight streaming from below the horizon - converge at the antisolar point on the opposite side of the sky, creating a pinkish glow. As the Sun sinks lower, the antisolar point rises higher and below it rises an arc of darker sky - Earth's shadow.

Glowing in the dark From dark sites, after total nightfall, you may be able to see one of the night sky's most elusive phenomena - the gegenschein. This very faint patch of light has the same origin as the band of zodiacal light that surrounds the Sun. It is centred on the ecliptic, but always lies directly opposite the Sun - it is caused by dust particles in the plane of the Solar System dimly reflecting back sunlight. The best times to see it are on moonless nights when it is well away from the glow of the Milky Way - October and February in the northern hemisphere, and April and August south of the equator.

46 Atmospheric phenomena

Our planet is surrounded by a life-giving envelope of gases – Earth's atmosphere. Although it is invisible, it scatters much of the visible light that arrives from space, creating many natural special effects in the process.

Earth's atmosphere is composed largely of nitrogen (78 per cent) and oxygen (21 per cent). The remaining 1 per cent is made up of inert gases, water vapour and a small but vital 0.03 per cent of carbon dioxide. These different gases interact with the earth, oceans and living organisms to create and maintain an environment suitable for life.

Although these gases seem transparent, the atmosphere has a tendency to diffuse light – an effect called scattering. Short-wavelength blue light interacts with the similar-sized molecules of gas and bounces off in random directions. This light arrives at our eyes from all directions, creating a general blue haze in the sky.

Scattering is most intense in the daylight sky away from the Sun, and can be seen over just a few hundred metres – the aerial perspective effect. At night, light sources are much dimmer – the Moon and stars – but scattering is still strong enough to turn the sky a deep, velvet blue.

Larger particles in the atmosphere – dust, water droplets and ice crystals – create other optical effects by refraction, diffraction and simple reflection. Clouds, for example, can produce dazzling effects. Made of water droplets suspended in the air, clouds may appear light, dark or in-between depending on the size of the droplets. Small ones scatter all colours of light and appear white in sunlight, whereas larger ones absorb the light, and appear grey.

Around the edges of clouds, colours often change as diffraction by the water droplets bends sunlight toward the observer. Diffraction has different effects on different colours of light, so clouds often have iridescent or mother-of-pearl fringes around their edges. When the Sun lies behind a thin veil of clouds, it is possible for the colours to extend all the way across them.

Rainbow colours A rainbow is a combination of refraction and reflection in raindrops. Each drop acts as a tiny prism, bending different colours of light by different amounts, while its inner surface acts as a mirror to bounce the light back toward an observer. A rainbow always forms at a fixed distance from the antisolar point. A faint secondary bow can form when light is reflected twice inside some water droplets.

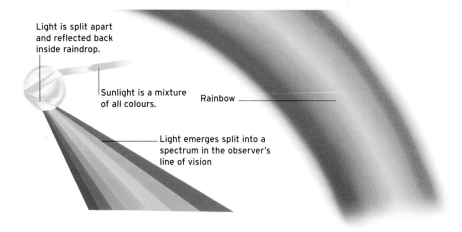

Light is split apart and reflected back inside raindrop.

Sunlight is a mixture of all colours.

Rainbow

Light emerges split into a spectrum in the observer's line of vision

See also: Our star, the Sun 36 · The solar cycle 38 · The wind from the Sun 42 · Earth – the dynamic planet 44

Special effects

Icy beauties *Many of the most beautiful sky effects are caused by ice crystals acting as tiny prisms and refracting light in distinctive ways. They are most spectacular around the Sun, but can also be seen, perhaps more easily, surrounding the Moon. Water can freeze into ice crystals with many different shapes, ranging from simple hexagonal drums to multifaceted diamonds.*

Glowing halo The most common type of halo occurs when light from the Sun or Moon is refracted by hexagon-shaped crystals and forms a ring of light at an angle of 22 degrees from the light source.

Sundogs Ice crystals forming the 22-degree halo refract light most effectively toward an observer when they appear level with the Sun in the sky. Here, they can produce bright sundogs, which may split sunlight as they refract it, taking on a rainbowlike appearance.

Sun pillars Great columns of light can sometimes be seen extending into the sky above the sunset. They are caused by flat ice crystals raining down from a cloud and reflecting the sunlight as they fall.

Noctilucent clouds Another phenomenon caused by particles in the atmosphere are these glowing clouds. At around 85 km (55 miles) up, they are far above normal weather systems. They seem to form in the coldest levels of Earth's atmosphere, toward the pole, around midnight. The ice crystals that form them must be condensing around a seed but, at this height, dust should be virtually absent from the atmosphere. One theory is that the clouds are seeded by meteor dust sifting down from space.

48 Meteors

On any dark cloudless night, you are likely to see shooting stars flashing across the sky. They are short-lived and vanish almost immediately. Yet many times each year shooting stars arrive in predictable showers that can sometimes be spectacular.

Showers of light Dust particles in a meteor stream are all locked in similar orbits around the Sun, so they rain down on Earth from the same direction. But due to the effects of perspective the meteors appear to diverge from a single point, called the radiant.

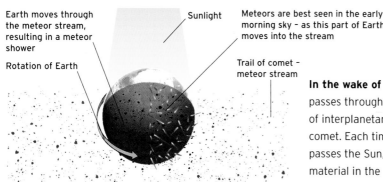

Earth moves through the meteor stream, resulting in a meteor shower

Rotation of Earth

Sunlight

Meteors are best seen in the early morning sky – as this part of Earth moves into the stream

Trail of comet – meteor stream

In the wake of a comet Earth passes through a dense band of interplanetary dust left by a comet. Each time the comet passes the Sun, it deposits new material in the meteor stream.

A shooting star or meteor is caused by a tiny particle of dust entering Earth's atmosphere. As it plunges downward, it collides with air molecules, and the friction generated causes the particle to glow and eventually burn up completely. Most meteors burn up at altitudes of around 100 km (60 miles). A few larger ones become fireballs – shooting stars that outshine all the stars in the sky. The inner Solar System is full of dust grains and potential meteors drifting around the Sun in their own orbits, often in the wake of comets. When Earth meets meteors head on, they enter the atmosphere travelling at up to 70 km/s (45 mps) but, when a meteor approaches Earth from behind, the speed can be much lower.

The larger and slower a meteor is, the longer it will survive as it enters the atmosphere. At lower altitudes, the air is thick enough for the meteor to leave a visible smoke trail behind it and sometimes even make an audible sound. Bolides are slow-moving fireballs that drop to below the speed of sound – roughly 1,220 km/h (760 mph) – with a sonic boom.

Most meteors arrive in regular showers or even storms. Meteor storms are rarer and less predictable than showers. They happen when Earth crosses a comet's trail shortly after its passage around the Sun, and can result in vast numbers of meteors. However, the dust-rich parts of the meteor stream that create a storm are ususally very thin, and Earth may arrive just an hour too early or too late for the main event.

Intriguingly, large meteors – meteorites – may have been the originators of life on Earth. Many of the bodies in interplanetary space are rich in carbon-based organic chemicals which, carried to the hot, wet conditions of the early Earth, could have become the raw material for primitive life.

See also: Genesis 34 · Our star, the Sun 36 · The wind from the Sun 42 · Earth – the dynamic planet 44 · Comets 78

Watching shooting stars

Storm watch *The best way to watch for meteors is simply to find a dark site with a clear view in the direction of the radiant, and set yourself up with a comfortable chair, flashlight and sky map. The number of meteors you are likely to see depends on the sky conditions and the location of the radiant. The meteor rates given in the table below are zenithal rates – the number of meteors you might see in ideal conditions, with the radiant directly overhead. One annual shower that occasionally produces great storms is the Leonid shower of mid-November. Every 33 years, when Comet Tempel-Tuttle sweeps past the Sun, leaving a dense trail of dust in its wake, the Leonid rate can increase a thousandfold or more. But predicting when a shower will produce a storm is difficult – storms last just a few hours, and Earth may miss them completely, or the peak activity may occur over relatively uninhabited areas. In the case of the Leonids, the structure of the meteor stream produces a shower of larger debris capable of producing fireballs a few hours before the main storm.*

STAR STORIES

The Great Storm of 1833
On the night of November 12, 1833, the greatest meteor storm in recorded history rained down over North America. Thousands of shooting stars streaked across the sky every minute, and fireballs bright enough to cast shadows left lingering smoke trails behind. The storm made an unforgettable impression on all who saw it. The 1833 Leonids also inspired astronomers to take meteors seriously for the first time - until then, many had dismissed them as an atmospheric phenomenon. Soon, reports of a storm seen on November 12, 1799, were found, and people realized that there was a periodic cycle behind the Leonids. By 1866, the Leonid meteors were linked to the orbit of a comet.

MAJOR METEOR SHOWERS

NAME	CONSTELLATION	DATES	MAX HOURLY RATE
Quadrantids	Boötes	Jan 1–6 (Max: Jan 3)	60
Lyrids	Lyra	April 19–25 (Max: April 22)	10
Eta Aquarids	Aquarius	May 1–10 (Max: May 6)	35
Delta Aquarids	Aquarius	July 15–Aug 15 (Max: July 29, Aug 7)	20, 10
Perseids	Perseus	July 23–Aug 20 (Max: Aug 12)	75
Orionids	Orion	Oct 16–27 (Max: Oct 22)	25
Taurids	Taurus	Oct 20–Nov 30 (Max: Nov 5)	10
Leonids	Leo	Nov 15–20 (Max: Nov 17)	10
Geminids	Gemini	Dec 7–15 (Max: Dec 13)	75

50 The Moon

Earth is not alone in its annual journey around the Sun – the Moon travels with us, orbiting the planet once every 27.3 days. Our Moon is one of the largest satellites in the Solar System and preserves a rocky record long since vanished from Earth.

A chip off the old block Astronomers today think that our satellite formed from material knocked off the early Earth shortly after its formation. Earth collided with another young world about the size of Mars. The other planet was destroyed and a large amount of material from it and from Earth's crust was flung into orbit around Earth.

The Moon still holds many mysteries despite being our nearest neighbour in space, and the only other world visited by humans. The most profound is its origin. Astronomers once held that it formed from the same cloud of gas and dust as Earth, or that it originated elsewhere in the Solar System and was captured by Earth's gravity. The rocks brought back by the Apollo lunar missions challenged both these theories. Moon rocks are too similar to those from Earth to have originated in a different part of the Solar System, but show important differences in composition. Furthermore, captured moons tend to be much smaller than their planets.

An important clue to the Moon's origin comes from the tides. Over billions of years, they have slowed the Moon's rotation so that it now spins on its axis in the same time that it orbits the Earth. The Earth's rotation is also slowing down and both will eventually be locked with just one face toward each other. The resulting forces are sending the Moon into a slow outward spiral, increasing its distance from Earth by 4 cm (1.6 in) every year.

Since the Moon is getting farther away from Earth, it must once have been much closer to it, and astronomers now believe it probably originated from an encounter, some 4.5 billion years ago, when Earth was involved in a collision with a large planetesimal.

MOON PROFILE

Visual magnitude (avg.)	-12.7	Equatorial diameter	3,474 km	
Distance from Earth (avg.)	384,400 km		2,159 miles	
	238,900 miles	Mass (Earth = 1)	0.012	
	60.3 Earth radii	Gravity at equator (Earth = 1)	0.17	
Orbital period	27.32 Earth days	Surface temperature	-150 to 120 °C	
Rotation period	27.32 Earth days		-240 to 240 °F	

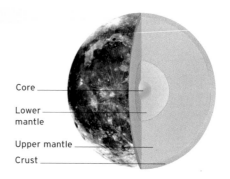

Core

Lower mantle

Upper mantle

Crust

See also: Earth – the dynamic planet 44 · The features of the Moon 52 · Probes to the planets 202

Our changing satellite

Phases of the Moon *The Moon's phases are caused by the changing illumination of its visible half as it circles Earth every 27.3 days. As the Moon orbits around Earth every month, sunlight illuminates different parts of the lunar globe. At New Moon, it is midday on the far side of the Moon, and midnight on the side facing Earth, so we cannot see the Moon at all. Then, as the Moon moves on, we see a sliver of light appear, growing or waxing toward a half-moon called First Quarter. Roughly 14 days after New Moon, the Moon lies directly opposite the Sun in the sky, and sunlight floods its Earth-facing side. This is the Full Moon. Then the Moon wanes as lunar night appears on its eastern edge, with the phase shrinking to Last Quarter. Finally, the Moon catches up with the Sun again at another New Moon, 29 days later.*

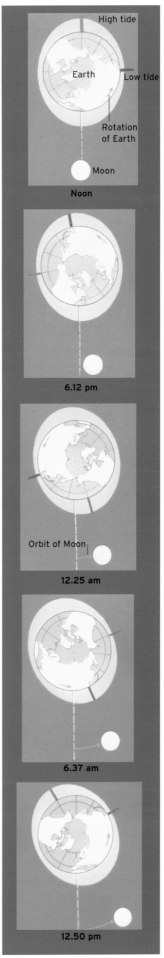

High tide
Earth
Low tide
Rotation of Earth
Moon
Noon

6.12 pm

Orbit of Moon
12.25 am

6.37 am

12.50 pm

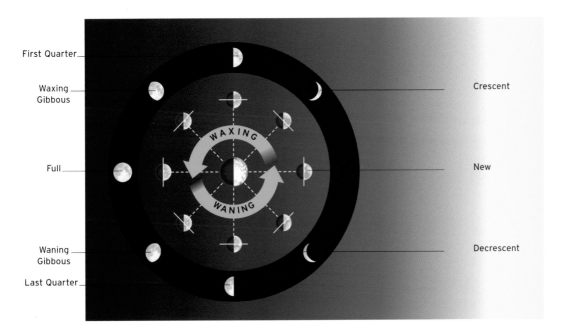

First Quarter

Waxing Gibbous

Full

Waning Gibbous

Last Quarter

WAXING

WANING

Crescent

New

Decrescent

Lunar eclipse If the Moon passes through Earth's shadow when on the opposite side of Earth from the Sun, a lunar eclipse occurs. However, the Moon does not disappear entirely from the sky when it is eclipsed. Instead it turns a dark orange or copper colour, as it reflects the faint glow of sunlight that filters through Earth's atmosphere.

Tidal movement The Moon creates tides as its gravity pulls on Earth's seas, causing a large bulge in ocean waters that is counterbalanced by a similar bulge on the opposite side of the planet. In effect, the Earth spins underneath these bulges, so that the sea levels around any coastline rise and fall twice in roughly 24 hours. The Sun has a similar effect, creating its own, smaller, tidal bulges. When the Sun and Moon are directly opposite each other, at New Moon or Full Moon, the bulges fall on top of each other, creating exceptionally high and low spring tides. At First and Last Quarter, the bulges tend to cancel each other out, leading to smaller neap tides.

Penumbra

Earth shadow (umbra) eclipsing Moon

Moon   Earth   Sun

52 The features of the Moon

Seen from Earth, the Moon clearly has two different types of terrain – bright highlands and dark seas of solidified lava. By studying its terrain, astronomers have unlocked the story of the Moon and found a key to the history of the Solar System.

A few decades ago, astronomers were divided over whether the Moon's craters were formed by meteor impacts or volcanoes. Only a few impact craters were known on Earth, and many doubted whether impacts alone could have covered the entire lunar surface. The advent of spaceprobes resolved the matter easily, showing that the Moon's surface had literally millions of craters, many far too small to be volcanic so they had to be meteoric. Once this was established, it became fairly easy to work out how different areas of the surface evolved. Each crater had to be more recent than the surface on which it formed – so small craters overlaying large ones must be younger.

This simple discovery allowed astronomers to work out a rough history of the Moon: the sparse craters scattered across the lunar seas formed after the volcanic eruptions that created the seas themselves, the seas are younger than the basins they formed in, and these are younger than the highlands around them.

Astronomers used to think that cratering had happened at a steady rate throughout the history of the Solar System, but rocks brought back by the Apollo astronauts held some surprises. They proved that the lunar seas were 3 to 4 billion years old – so most of the cratering must have happened soon after the Moon's formation.

It now seems that most of the debris from the formation of the Solar System had been swept up by around 3.8 billion years ago, culminating in a series of vast collisions as the final planetesimals were soaked up, forming huge impact basins in which the lunar seas then formed. Since the lava seas solidified, such large impacts have been much rarer, although today a steady rain of micrometeorites continues to pulverize the Moon.

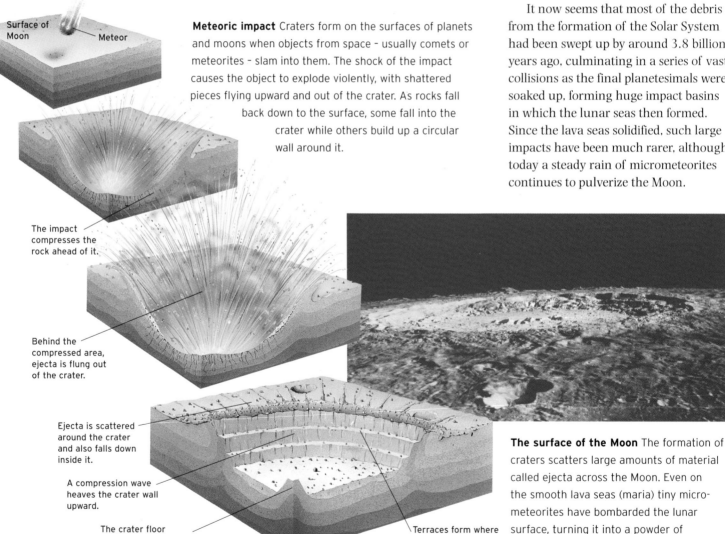

Surface of Moon

Meteor

The impact compresses the rock ahead of it.

Behind the compressed area, ejecta is flung out of the crater.

Ejecta is scattered around the crater and also falls down inside it.

A compression wave heaves the crater wall upward.

The crater floor rebounds in the middle, forming a central peak.

Terraces form where the crater walls slump downward.

Meteoric impact Craters form on the surfaces of planets and moons when objects from space - usually comets or meteorites - slam into them. The shock of the impact causes the object to explode violently, with shattered pieces flying upward and out of the crater. As rocks fall back down to the surface, some fall into the crater while others build up a circular wall around it.

The surface of the Moon The formation of craters scatters large amounts of material called ejecta across the Moon. Even on the smooth lava seas (maria) tiny micro-meteorites have bombarded the lunar surface, turning it into a powder of crushed rock, or regolith.

See also: The home system 32 · Genesis 34 · Meteors 48 · The Moon 50 · Probes to the planets 202

The face of our Moon

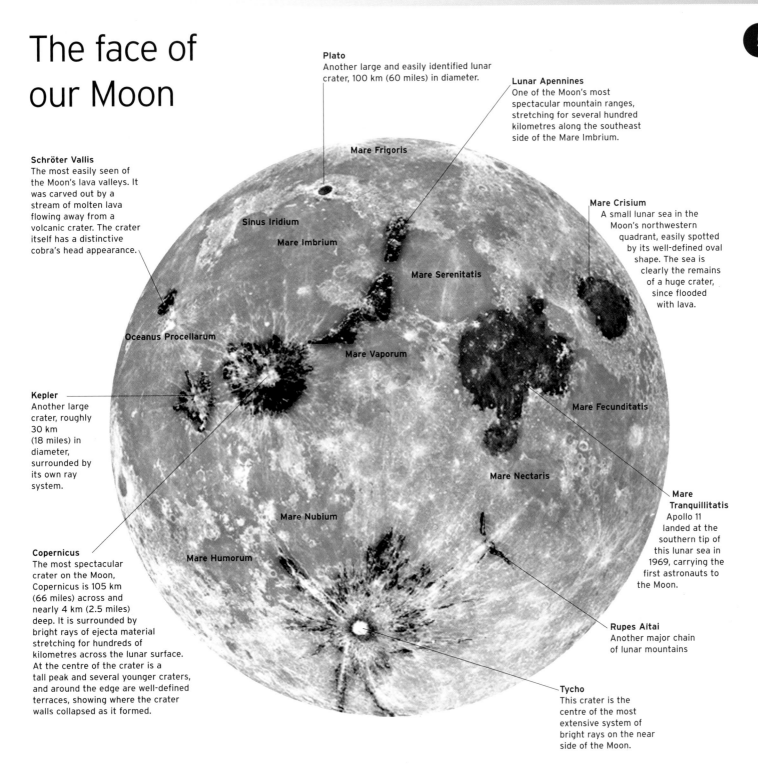

Plato
Another large and easily identified lunar crater, 100 km (60 miles) in diameter.

Lunar Apennines
One of the Moon's most spectacular mountain ranges, stretching for several hundred kilometres along the southeast side of the Mare Imbrium.

Mare Frigoris

Schröter Vallis
The most easily seen of the Moon's lava valleys. It was carved out by a stream of molten lava flowing away from a volcanic crater. The crater itself has a distinctive cobra's head appearance.

Sinus Iridium

Mare Imbrium

Mare Serenitatis

Mare Crisium
A small lunar sea in the Moon's northwestern quadrant, easily spotted by its well-defined oval shape. The sea is clearly the remains of a huge crater, since flooded with lava.

Oceanus Procellarum

Mare Vaporum

Kepler
Another large crater, roughly 30 km (18 miles) in diameter, surrounded by its own ray system.

Mare Fecunditatis

Mare Nectaris

Copernicus
The most spectacular crater on the Moon, Copernicus is 105 km (66 miles) across and nearly 4 km (2.5 miles) deep. It is surrounded by bright rays of ejecta material stretching for hundreds of kilometres across the lunar surface. At the centre of the crater is a tall peak and several younger craters, and around the edge are well-defined terraces, showing where the crater walls collapsed as it formed.

Mare Nubium

Mare Humorum

Mare Tranquillitatis
Apollo 11 landed at the southern tip of this lunar sea in 1969, carrying the first astronauts to the Moon.

Rupes Altai
Another major chain of lunar mountains

Tycho
This crater is the centre of the most extensive system of bright rays on the near side of the Moon.

Craters, mountains and seas *The Moon's features are as diverse as those of Earth, and many of them are visible using binoculars or a small telescope. They range from bright craters to seas, and from mountain chains to volcanic valleys. The appearance of such features can change significantly from day to day, but most are best seen around First and Last Quarter, when the Sun is low on the horizon and raised features cast long shadows.*

STAR STORIES

The Hare in the Moon

Many cultures have seen figures on the Moon's surface. In the west, the most common figure is the Man in the Moon, but the ancient Chinese saw the figures of a woman and a hare. According to legend, the woman is the beautiful Ch'ang-o, wife of the celestial archer I, who had been granted an elixir of immortality for saving the world. One day, Ch'ang-o drank the elixir and, when I found out, he chased her all the way to the Moon. There, the lunar hare gave Ch'ang-o protection, and she still lives there to this day.

54 Mercury–messenger of the gods

Mercury is the closest planet to the Sun and the smallest apart from Pluto. At first, it appears to resemble the Moon, but Mercury has many unique features – as do all the planets in the Solar System. It has a distinct history and personality.

Mercury's most impressive feature is the Caloris Basin – a crater over 1,300 km (800 miles) across, which flooded with lava at some time after its creation. It is the second-largest known crater in the Solar System, after the Aitken Basin on the Moon's far side, and its formation sent shockwaves all around the planet.

Mercury's craters differ from the Moon's because gravity is higher. This means that ejecta from an impact falls back much closer to the crater rim, instead of forming an extensive system of rays. The unusually high gravity is the planet's greatest mystery – it is three-quarters the size of Mars, yet has the same surface gravity. This would be explained if the planet contained an abnormally large amount of heavy material – an oversize iron and nickel core, for instance. Astronomers now suspect that the young Mercury suffered an enormous collision that blew away much of it's outer mantle and crust.

Craters and plains on Mercury are frequently cut by towering cliffs or steep-sided valleys that run for miles across the surface. These unique features suggest that, early in its history, the planet heated up, expanded and cracked its surface. When it cooled and shrank again, parts of its surface were pushed up and over adjacent areas.

Mercury's day is two-thirds of its year, but day and year combine together so that, for most of the planet, there is just one sunrise every two years. For other areas, sunrise coincides with Mercury's closest approach to the Sun, and highest speed in its orbit. At times like this, Mercury's movement can outpace its rotation, so the Sun sinks back toward the horizon, sets, and rises again in the course of a few days.

Mercury orbits the Sun bolt upright, and as a result the planet's poles only ever see the Sun close to the horizon. This means that deep craters in these areas are kept in permanent darkness and may shelter large ice deposits dumped by comets that have collided with Mercury.

The landscape of Mercury Mercury's surface, like the Moon's, is divided into cratered highlands and low, smooth plains. They tell a similar story to the Moon's own – one of heavy early bombardment and extensive lava flooding over broad regions.

PLANET PROFILE

Visual magnitude (avg.)	0.0		Equatorial diameter	4,875 km	
Distance from Sun (avg.)	57.9 million km			3,029 miles	Mantle
	36.0 million miles		Mass (Earth = 1)	0.055	Core
	0.387 AU		Gravity at equator (Earth = 1)	0.38	Crust
Length of year	88 Earth days		Surface temperature	-170 to 430 °C	
Rotation period	58.6 Earth days			-280 to 800 °F	

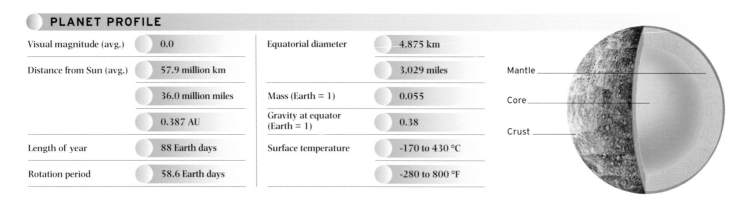

See also: The home system 32 · Genesis 34 · Meteors 48 · Venus – the hostile planet 56 · Probes to the planets 202

Finding and observing Mercury

Mercury locater chart *Mercury is the most difficult of the nearby planets to spot, primarily because of its position in the sky. The planet's orbit is small and we look at it from outside, so Mercury is only visible when it is very close to the Sun. At its maximum elongations – the turning points of its orbit as seen from Earth – the small planet can reach 28 degrees from the Sun. But this still leaves it trapped in twilight and, as the sky gets darker, Mercury sinks closer to the horizon, eventually disappearing behind a thicker and more turbulent layer of atmosphere.*

At its best, Mercury can be seen from Earth for just a few days either side of elongation, and its magnitude at these times can rival all but the brightest stars. The most successful way to spot Mercury is to use binoculars to scan the twilight sky just above the horizon near the point where the Sun has set. For best results, wait about 15 minutes after sunset to let the sky darken.

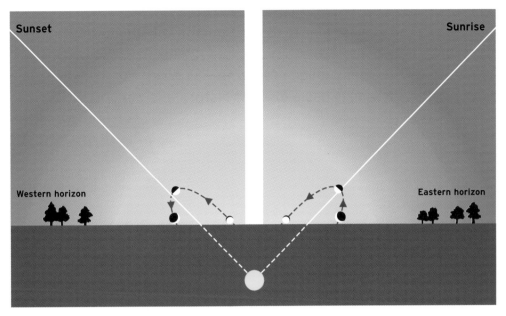

Eastern elongations
Sept. 2001	Feb, June, Oct 2006
Jan, May, Sept, Dec 2002	Feb, June, Oct 3007
April, Aug, Dec 2003	Jan, May, Sept 2008
March, July, Nov 2004	Jan, May, Aug, Dec. 2009
March, July, Nov 2005	April, Aug, Dec 2010

Western elongations
Oct 2001	April, Aug, Nov 2006
Feb, June, Oct 2002	March, July, Nov 2007
Feb, June, Sept 2003	March, July, Oct 2008
Jan, May, Sept, Dec. 2004	Feb, June, Oct 2009
April, Aug, Dec 2005	Jan, May, Sept 2010

Chance in a million Given the relatively rare occasions on which Mercury is visible from Earth, it is even rarer to have the moment caught on film. This amateur skyshooter caught Mercury soon after sunset, before it disappeared below the horizon. The elusive planet appears just above the sunset's dying glow.

✦ STAR STORIES

The Winged Messenger

The ancient Babylonians were the first people to associate swift-moving Mercury with their messenger-god, in this case, Nabu. Their influence on later civilizations led the Greeks to call the planet Hermes on its morning appearances and Apollo in the evening, even though they knew it was just one planet. Under the Romans, Hermes transformed into Mercury, the swift-footed messenger famous for his winged helmet. The Aztecs saw its appearances as mystical messengers from the underworld.

56 Venus – the hostile planet

Shrouded in white clouds, Venus is the second-brightest object in the night sky. Despite being named after the Roman goddess of beauty, Venus is the hottest and most deadly place in the Solar System.

Venus' atmosphere was a mystery until spaceprobes revealed that the brilliant clouds were quite unlike the water-vapour clouds of Earth, but were instead made of sulphuric acid and other deadly and corrosive compounds. In fact, most of Venus' atmosphere is made up of carbon dioxide, with no sign of the nitrogen and oxygen that make life on Earth possible. At the planet's surface, the temperature rises to a searing 470 °C (870 °F) due to the trapped solar radiation that cannot escape the thick clouds, and atmospheric pressure is nearly 100 times greater than on Earth.

Venus has no water either: the carbon dioxide overload long ago created a runaway greenhouse effect that boiled any liquid water on the Venusian surface. Fierce sunlight then split the water molecules into hydrogen and oxygen, and the hydrogen was lost to space while the oxygen was locked up in yet more carbon dioxide.

The lack of water on Venus has given the planet a completely different geology and history from Earth. Water helps to weather rocks, turning them into lighter carbonate minerals and absorbing carbon dioxide from the

Hidden truth
For a long time, Venus' cloud covering prevented us from discovering the planet's most basic details such as its rotation period. This Magellan probe image (taken with cloud-penetrating radar) shows the surface of Venus stripped of cloud.

atmosphere in the process. It also contributes to volcanic activity, which releases heat from our planet's interior.

Earth and Venus are roughly the same size and probably generate the same amount of heat inside them. However, on Venus the heat cannot escape at a steady rate, so it builds up like a pressure cooker until planet-wide volcanic eruptions re-cover the entire surface in just a few million years. Radar maps of the surface show it covered in

lava flows, with just a few impact craters scattered across the surface. Assuming Venus suffers impacts at roughly the same rate as Earth, it seems the entire planet was last resurfaced around 500 million years ago.

There is no conclusive evidence to show whether Venus is currently active volcanically. Recent measurements of atmospheric gases, however, show changes that suggest volcanoes may still erupt on Venus today.

PLANET PROFILE

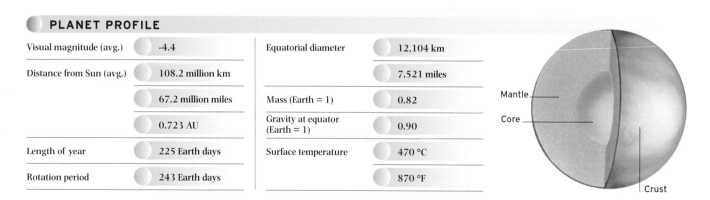

Visual magnitude (avg.)	-4.4	Equatorial diameter	12,104 km	
Distance from Sun (avg.)	108.2 million km		7,521 miles	
	67.2 million miles	Mass (Earth = 1)	0.82	
	0.723 AU	Gravity at equator (Earth = 1)	0.90	
Length of year	225 Earth days	Surface temperature	470 °C	
Rotation period	243 Earth days		870 °F	

Mantle

Core

Crust

See also: The home system 32 · Genesis 34 · Earth – the dynamic planet 44 · Probes to the planets 202

Finding and observing Venus

Venus locater chart *Venus, like Mercury, is closer to the Sun than Earth and is only ever seen close to the Sun. However, its orbit is much wider and closer to us than Mercury's, so the planet can get much farther away from the Sun in twilight, and is the brightest object in the night sky, except for the Moon. The Venus locater shows how to find Venus between 2001 and 2010. The loops show the path of Venus through the evening sky (east of the Sun), and the morning sky (west of the Sun). Dates of greatest elongation are shown for each apparition, but Venus is visible for several months on either side of these dates.*

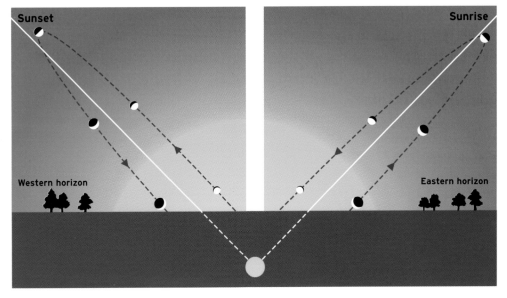

Eastern elongations	
Feb 2002	Dec 2006
Sept 2003	July 2008
Jan 2004	Jan 2009
May 2005	Feb 2010

Western elongations	
Dec 2002	Sept 2007
July 2004	Jan 2008
Jan 2005	April 2009
Feb 2006	Dec 2010

Shining bright When seen from Earth, Venus is at its brightest shortly before it disappears from the evening sky and passes into the morning. Venus spins very slowly - its day is longer than its year - and it is tilted so far over in its orbit—177.3 degrees—that it spins in the opposite direction to other planets.

Venus in crescent Venus' phase is caused in the same way as the Moon's: different amounts of the planet's sunlit side are presented to Earth depending on the planets' relative positions. The phase is most obvious when Venus is close to Earth, shortly before and after it passes between us and the Sun. The thin crescent is visible even in low-powered binoculars, and sharp-eyed observers may be able to see it with the unaided eye.

⊕ **STAR STORIES**

Venus in Mesoamerica

To the ancients, Venus' brilliance meant it was associated with goddesses of love and beauty, but ancient civilizations of Central America had a very different view. Both the Maya and Aztec peoples developed calendars and elaborate rituals based on the cycle of Venus' movements. The early Mayan civilization associated Venus with Tlaloc, the god of disaster, and even planned their warfare around the planet's movement. Later, the planet became a supreme god - the feathered serpent Kukulcán.

58 Mars - the Red Planet

Of all the planets in the Solar System, Mars is the one most like Earth. The possibility of finding life – or traces of former life – on Mars has compelled much study of the Red Planet, but the question still remains unanswered.

Mars is the last of the rocky inner planets before the asteroid belt and the giant outer worlds. It is just over half the diameter of Earth, and so it has cooled much more than Earth since its formation. However, early in its history the planet must have been an active world. Like Venus, Mars shows no sign of continental plates, and with its surface locked solid, heat had to escape the planet through massive volcanoes.

The Tharsis Bulge, a volcanic plateau the size of the United States, stands 10 km (6 miles) above the average Martian surface, and is topped by three huge volcanoes. A network of linear canyons—Valles Marineris—which runs 4,000 km (2,500 miles) and reaches depths of 6 km (3.7 miles), formed as the Martian surface gave way when the Tharsis Bulge lifted up. Nearby stands Olympus Mons, a volcano 21 km (13 miles) high – nearly three times higher than Mount Everest. The surfaces of these volcanoes are relatively young and uncratered, indicating they were probably active as recently as 150 million years ago – too recent to judge whether they are truly extinct today.

The Martian terrain roughly divides into two halves. The northern hemisphere is covered by low-lying smooth plains, whereas the south stands higher and is heavily cratered. One intriguing idea is that Mars once had an ocean or several lakes filling depressions in the northern lowlands. These may have harboured life before evaporating or draining away, and traces may still be found in layers of sediments there.

PLANET PROFILE

Visual magnitude (avg.)	-2.0
Distance from Sun (avg.)	227.9 million km
	141.6 million miles
	5.203 AU
Length of year	1.88 Earth years
Rotation period	24.62 Earth hours
Equatorial diameter	6,780 km
	4,213 miles
Mass (Earth = 1)	0.11
Gravity at equator (Earth = 1)	0.38
Surface temperature	-130 to 30 °C
	-210 to 80 °F

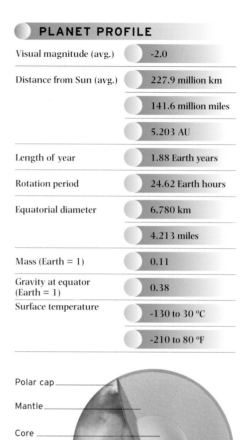

Polar cap
Mantle
Core
Crust
Surface

Red eye effect Mars' distinctive reddish-brown colour is caused by large amounts of iron oxide (rust) in the soil. Through the Hubble Space Telescope, the polar caps can be seen clearly and, although Mars' atmospheric pressure is less than 1 per cent of Earth's, faint wisps of cloud are still visible. What air there is on Mars is 95 per cent carbon dioxide, with small amounts of nitrogen and argon, and only traces of other gases such as water vapour.

See also: The home system 32 · Genesis 34 · On Martian soil 60 · Probes to the planets 202

Finding and observing Mars

Mars locater chart *Mars is easy to spot in the night sky because of its obvious red colour. Over the course of its orbit, its distance from Earth and brightness in the sky change rapidly. When Mars is in opposition – at its closest approach to Earth – it can reach magnitude -2.8. However, when it is in conjunction – at its most distant – the Red Planet can fade to magnitude -1.*

The Mars locater allows you to work out the approximate position of Mars at any time up to 2006. The outer ring shows Mars' changing position in the zodiac. The inner ring shows Earth's changing position through the year. Where the line from Earth to Mars passes to the right of the Sun, Mars can be seen rising before the Sun in the pre-dawn sky. If the line passes to the left, then it sets after the Sun in the evening sky. Where the line passes nowhere near to the Sun, Mars is in opposition and is visible all night. Bright oppositions of Mars occur once every 2 to 3 years, with the next one in August 2003. As Mars' orbit is noticeably elliptical, every 17 years the oppositions are much closer than others and are much brighter.

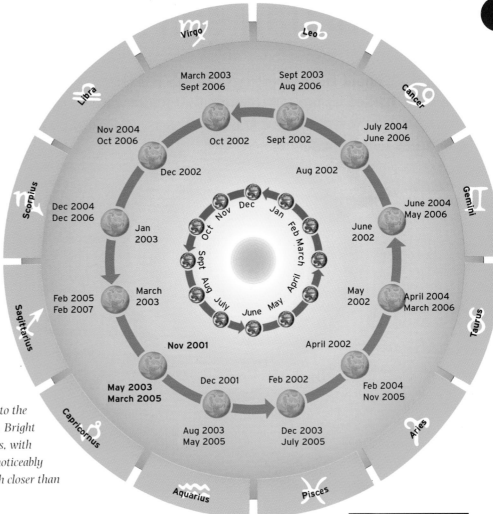

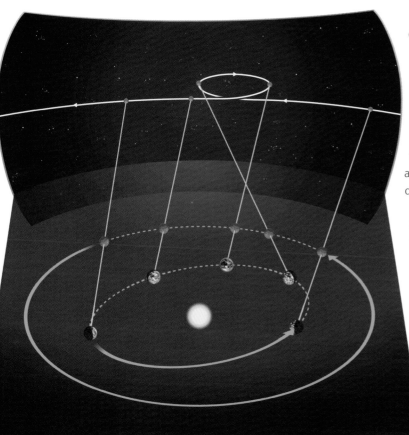

Coloured filters Mars is a small planet, and viewing its surface features with a telescope is a challenge. Coloured filters can help by increasing the contrast of certain features. A yellow filter darkens the bluer features - lava flows and boulder fields - whereas a green filter allows the white polar caps and cloud features to stand out.

Rewind Earth travels faster than Mars as it has a smaller orbit. When Earth is closest to Mars, it overtakes the Red Planet, and Mars appears to travel backward in the sky. This effect, called "retrograde motion", affects all the outer planets, but is most obvious for Mars. For several months, Mars makes a large loop through the zodiac constellations before it moves forward again.

MARS UNFILTERED

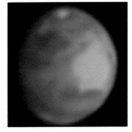

YELLOW FILTER

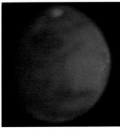

GREEN FILTER

60 On Martian soil

Arguments about the possibility of simple life forms on Mars continue, fuelled by new discoveries that liquid water forms for short periods on the surface and may also be present in larger quantities below the planet's surface.

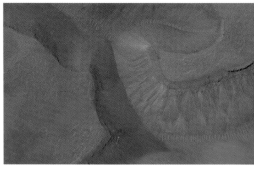

Surface water on Mars In the Gorgonum Chaos region of Mars, gullies have been found that cut into the edges of a steep-sided canyon. As these are reminiscent of water features on Earth, they have been taken as evidence that water occasionally breaks the surface on Mars. It is thought that flash floods spring from underground reservoirs before the water boils away in Mars' low-pressure atmosphere.

Life on Mars? Speculation ran wild in 1996 when a team of NASA scientists reported traces of life in a chunk of Martian rock that had arrived on Earth as a meteorite. Their discoveries were controversial – they claimed that certain chemicals in the rock could only have been created by life, and even reported possible fossilized bacteria.

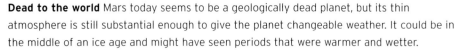

Dead to the world Mars today seems to be a geologically dead planet, but its thin atmosphere is still substantial enough to give the planet changeable weather. It could be in the middle of an ice age and might have seen periods that were warmer and wetter.

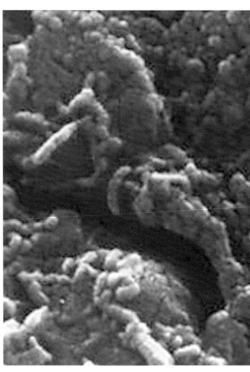

Mars, like Earth, has seasons. The planet is tilted just slightly more than Earth and experiences a similar pattern of changes as first the north and then the south pole points toward the Sun over the course of a Martian year. The changing temperatures cause carbon-dioxide ice from the Sunward-facing polar cap to evaporate into the atmosphere, while a new layer of carbon-dioxide frost condenses from the atmosphere onto the winter cap. Although the Martian polar caps seem similar, orbiting spaceprobes have discovered that they are quite different. Each is built up in annual layers a fraction of a millimetre thick, forming an unchanging cap of permafrost, surrounded by a broader expanse of seasonal frost. The southern icecap seems to be entirely carbon dioxide, but the northern cap is probably underlaid with a permanent layer of frozen water.

Evidence for water below the surface of Mars has been steadily mounting since the 1970s. The Mariner and Viking probes photographed broad channels across the northern plains that might have been caused by flash floods, and narrower winding valleys that may have been cut by long-dried-up rivers. More recent images show sediments laid down by geologically young watercourses and lake beds.

See also: The home system 32 · Genesis 34 · Mars – the Red Planet 58 · Probes to the planets 202

Martian seasons

Poles apart *The polar caps are probably the easiest features to see on the Martian surface, and it is interesting to track their changes from season to season to see how they differ from each other. The northern cap, pictured here in winter, is underlaid with substantial deposits of water ice, so only varies a little in size over the Martian year – helped by the fact that northern summer happens when Mars is farthest from the Sun. The southern cap, however, lacks such deposits of water ice and varies dramatically in size. It can all but vanish in the hot summer, but regenerates rapidly in the extremely cold winter.*

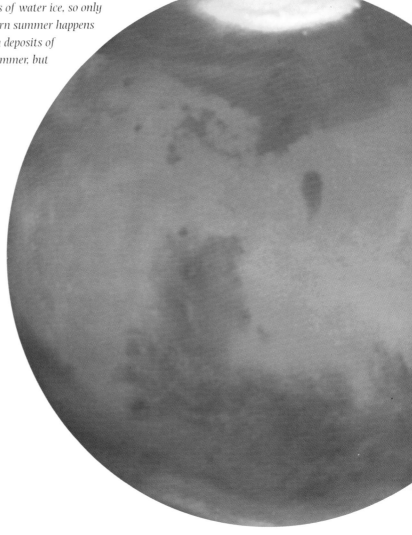

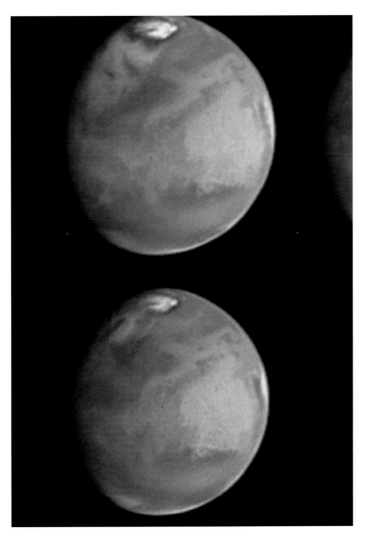

Stormy weather Mars' elliptical orbit comes closest to the Sun in southern spring and summer. High winds blow from the warm south pole into the northern hemisphere, stirring up fine dust and creating vast storms that can engulf large areas of the planet.

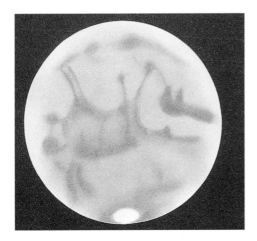

Distinguishing features Using a small telescope, it is possible to discern several features that stand out in contrast with their backgrounds. In general the Martian highlands of the southern hemisphere are darker than the planet's northern plains. Easy to spot are the dark, cratered volcanic plains such as the Mare Acidalium and Syrtis Major in the north, and the large, bright, impact craters Hellas and Argyre in the south. Filters can help increase the contrast further, picking out early morning frosts on the crater floors.

 STAR STORIES

Canals of Mars

In 1877, Italian astronomer Giovanni Schiaparelli interpreted fine lines running between the dark Martian plains as natural channels carrying water from the polar caps to areas of vegetation, but a mistranslation turned them into canals, implying they were the product of intelligent life. For years afterward, some astronomers insisted they could see the canals, and the idea of vegetation on the Red Planet lasted until the beginning of the Space Age.

62 Asteroids

Between Mars and Jupiter lies the no-man's-land of the Solar System – a broad stretch of space inhabited only by chunks of rock. These asteroids, which never formed into a planet, litter space from Earth's orbit out as far as Saturn.

The asteroid belt between Mars and Jupiter contains hundreds of billions of fragments of rock that range from pebbles and boulders to worlds the size of small moons. These objects orbit in a zone 254 to 598 million km (158 to 372 million miles) from the Sun. At certain distances from the Sun, the belt has empty regions called "Kirkwood gaps". Any asteroid orbiting within these zones will find itself frequently aligned with Jupiter, and the giant planet's powerful gravity soon perturbs its orbit, catapulting it from the belt into a new and different orbit.

Astronomers used to think that the asteroid belt was the remnants of a tenth planet between Mars and Jupiter that was shattered to pieces by some cataclysmic event. However, it now seems that the gravitational attraction toward Jupiter was always too strong for a planet to coalesce so close to it.

Trapped within this belt the asteroids jostle and collide. A few of the largest planetesimals survive, but many of the others have smashed each other apart over billions of years. The largest surviving asteroid, Ceres, is a ball of rock 930 km (580 miles) across, and there are 50 others with diameters more than 160 km (100 miles) across. The vast majority of the asteroids are just several tens of kilometres wide or less. These smaller worlds often have bizarre shapes, as they lack the gravity to pull them even roughly into spheres. A few seem to be little more than orbiting piles of rubble held loosely together by mutual gravitational attraction.

Jupiter's gravity has scattered asteroids around the inner regions of the Solar System, many of them coming close to, and sometimes even crossing, Earth's orbit. Small rocky fragments in this sort of orbit are the source of most of the meteorites that land on Earth's surface. Jupiter also flung a few asteroids into longer orbits, or out of the Solar System altogether, while a number ended up sharing an orbit with the giant planet itself, leading or trailing it at a safe distance in two clusters known as the Trojans. This scattering is still going on today, as occasional collisions between asteroids nudge them into different, less stable orbits.

Asteroids are small, so they did not heat up a great deal during their formation and have therefore preserved material unaltered in the 4.6 billion years since the Solar System was created. Many asteroids seem to be linked to different types of meteorites – some seem to be pure metal, some are stony-iron and others are dark objects rich in carbon-based chemicals – some may even be dead comets. Just a few of the larger asteroids show evidence of volcanic minerals on their surface, indicating that they may have once had molten interiors.

Eros, Ida and Gaspra These typical asteroids all lack sufficient mass to pull themselves into proper spheres. Ida and Gaspra (right) are main-belt asteroids, around 32 km (20 miles) and 16 km (10 miles) long respectively. The number of craters on each hints at their complex history, going back millions or billions of years. The near-Earth asteroid Eros (left) appears to be an elongated single piece of rock, about 35 km (22 miles) long and 13 km (8 miles) wide.

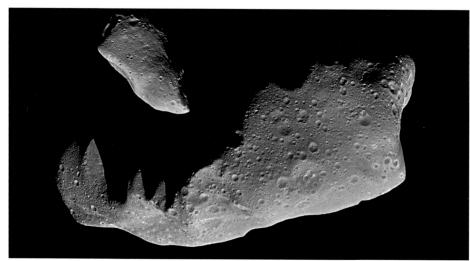

See also: The home system 32 · Genesis 34 · Meteors 48 · Astronomy from space 200

Vermin of the skies

Tracking bright asteroids *Ceres and Vesta shine at magnitudes 6.7 and 5.2 respectively, so they are only visible if at least one constellation separates them from the Sun. Even then, you will need a more detailed locater map to find them.*

The minor planet locater shows the movement of the bright asteroids Ceres and Vesta between 2001 and 2005. Both of these asteroids orbit roughly in the plane of the ecliptic, and spend most of their time within the constellations of the zodiac. To use the locater, simply find the positions of Earth and the asteroid for the current date. If the Sun lies to the right of this line, then the asteroid is visible in the evening sky after sunset. If the Sun lies to the left, it is visible in the morning before sunrise.

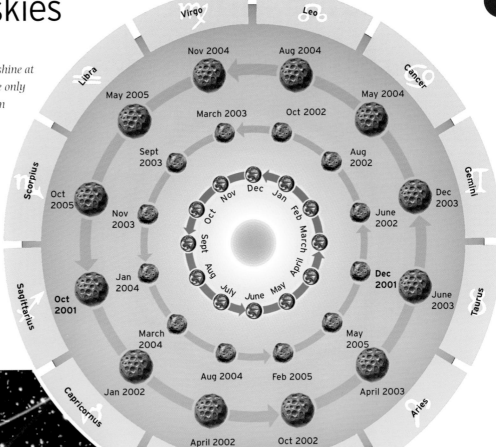

Say "cheese" Asteroids are called the vermin of the skies for their annoying habit of turning up in long-exposure photographs of distant objects. Here, a magnitude 18.7 asteroid (blue) passing through Centaurus was caught by the Hubble Space Telescope.

INTERESTING ASTEROIDS

NAME	NUMBER	TYPE	NOTES
Ceres	1	Main belt	First asteroid found, 1801
Pallas	2	Main belt	Second asteroid found, 1802
Juno	3	Main belt	Third asteroid found, 1804
Vesta	4	Main belt	Volcanic rock on surface?
Hektor	624	Trojan	Shares orbit with Jupiter
Amor	1221	Near-Earth Asteroid (NEA)	Orbit comes close to Earth's
Apollo	1862	NEA	Orbit mainly outside Earth's
Aten	2062	NEA	Orbit mainly within Earth's
Eros	433	NEA	Orbited by NEAR probe, 2000-1
Mathilde	253	Main belt	Visited by NEAR probe, 1997

64 Jupiter - king of the planets

Jupiter is a giant among worlds, large enough to contain all the other eight planets put together. But the king of the Solar System is mostly a vast and near-bottomless ocean of clouds, gas and liquid, surrounding a much smaller, rocky core.

A perfect storm Jupiter's Great Red Spot is a storm three times the size of Earth that has lashed the planet's surface for centuries. The spot is a high-pressure cloud that absorbs smaller storms driven into it by Jupiter's powerful east-west winds.

Jupiter is huge – Earth could easily be swallowed up in just one of the giant storms that rack the planet. It is also very light, made almost entirely of the gas hydrogen, a little helium and traces of other elements. The farther you fall into Jupiter, the greater the pressure gets – several hundred kilometres down, the weight from above is so huge that hydrogen behaves as a liquid.

Jupiter rotates rapidly and, as the interior swirls around, an intense magnetic field is generated whose influence reaches as far as the orbit of Saturn. The core of the planet is a ball of rock larger than Earth. The core's gravity causes the outer layers to contract, increasing the pressure and temperature inside so that Jupiter gives out more heat than it receives from the Sun. Some astronomers call Jupiter a failed star – if it had formed with more mass, it could have become hot enough to begin nuclear fusion and start shining on its own.

Jupiter's atmosphere is dominated by multicoloured streamers of blue, brown, cream and red clouds. The colours are caused by complex chemical reactions inside the clouds, based on the element sulphur. The planet's rapid spin stretches clouds out around the planet, forming broad belts blown by winds at hundreds of kilometres per hour in alternate directions. Where these belts and zones meet, gigantic storms brew.

PLANET PROFILE

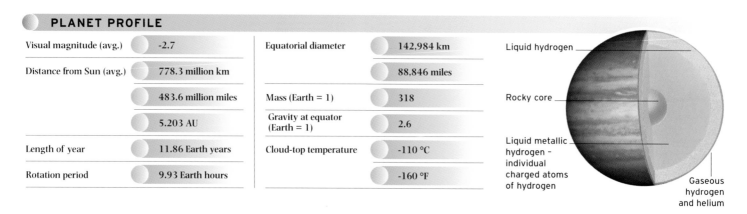

Visual magnitude (avg.)	-2.7	Equatorial diameter	142,984 km	Liquid hydrogen
Distance from Sun (avg.)	778.3 million km		88,846 miles	
	483.6 million miles	Mass (Earth = 1)	318	Rocky core
	5.203 AU	Gravity at equator (Earth = 1)	2.6	
Length of year	11.86 Earth years	Cloud-top temperature	-110 °C	Liquid metallic hydrogen – individual charged atoms of hydrogen
Rotation period	9.93 Earth hours		-160 °F	Gaseous hydrogen and helium

See also: The home system 32 · Genesis 34 · The Galilean moons 66 · Probes to the planets 202

Finding and observing Jupiter

Jupiter locater chart *Jupiter is one of the easiest planets to spot – even at its greatest distance from Earth, it is still brighter than any star, and has a whitish colour. Ordinary binoculars will reveal Jupiter's four brightest moons, looking much as they did when Galileo discovered them in 1610. A telescope shows Jupiter's disc, made oval by its rapid rotation. Telescopes also reveal dark belts and bright zones of slowly shifting clouds, which make up Jupiter's surface.*

The Jupiter locater allows you to work out Jupiter's position at any time up to 2012. The outer ring allows you to see which of the 12 zodiacal constellations Jupiter lies in, and the inner ring shows Earth's changing position throughout the year. Where the line from Earth to Jupiter passes to the right of the Sun, Jupiter can be seen rising before the Sun in the pre-dawn sky. If the line passes to the left, then it sets after the Sun in the evening sky.

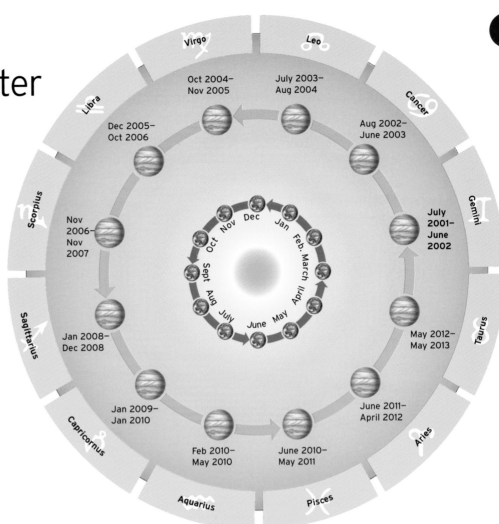

JUPITER UNFILTERED

BLUE FILTER

YELLOW FILTER

Coloured filters The visibility of Jupiter's cloud features will be enhanced by a colour filter held over a telescope eyepiece. Try a light blue filter to darken the Great Red Spot, and a reddish or yellow one to bring out details in the bands and polar regions. These images show how filters improve the view.

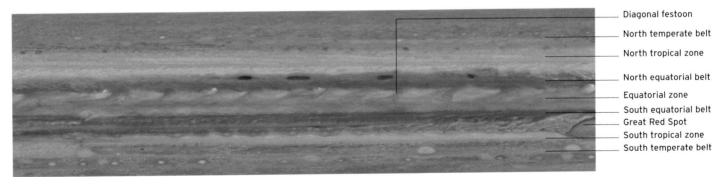

- Diagonal festoon
- North temperate belt
- North tropical zone
- North equatorial belt
- Equatorial zone
- South equatorial belt
- Great Red Spot
- South tropical zone
- South temperate belt

Jupiter's cloud belts and zones To study Jupiter in detail, it is helpful to know about the planet's different weather systems. Astronomers divide the cloud bands into light-coloured zones and darker belts. The belts and zones are formed of clouds at different heights in the atmosphere. Diagonal festoons and white storms often occur at the boundaries between them.

The Galilean moons

66

Jupiter is a huge planet with a family of at least 28 moons. Four of these – the Galilean satellites Io, Europa, Ganymede and Callisto – are spectacular worlds in their own right, each with unique features, such as erupting volcanoes and hidden oceans.

The moons of Jupiter number more than two dozen, and divide into several distinct groups. Four tiny inner satellites orbit just outside the planet's thin dust ring, and at least 20 more are probably captured asteroids that orbit millions of miles out from Jupiter itself. Between these groups lie the four Galilean moons that range in size from slightly smaller than our own Moon to 1.5 times its size.

Io is the closest Galilean moon to Jupiter. Trapped in a gravitational tug-of-war between the planet and the other satellites, its interior is constantly pulled this way and that, heating it up and generating the huge volcanoes that cover much of its surface. Io's most violent volcanoes are more like geysers, spraying plumes of liquid sulphur compounds into space. Elsewhere, volcanic lakes of molten sulphur glow with heat. These different forms of elemental sulphur colour the surface in a swirling palate of yellows, oranges and browns.

Second of the Galilean satellites from Jupiter, and just slightly smaller than Io, Europa is an enigma that could contain the Solar System's greatest secret – extraterrestrial life. This moon's surface looks forbidding, covered in brilliant ice and as smooth as a cue ball, but closer inspection reveals that the ice has cracked apart and healed itself many times, rather like an arctic ice floe. There is now mounting evidence for a global ocean of water beneath the frozen surface, kept liquid by the same tidal heating that affects Io. Here, complex life could have developed around undersea volcanic vents, just as it may have done on Earth.

Ganymede is the largest of the Galilean moons, and the third one out from Jupiter. Being larger, it should be hotter inside than Io and Europa, but Ganymede's distance from Jupiter means it does not benefit from tidal heating. The moon is covered in large dark regions, separated by areas of

lighter terrain with long grooves. Astronomers think that these terrains reveal a history of moving plates similar to those on Earth. Early in Ganymede's history, its own internal heat would have been enough to melt it from the inside out. The dark, early crust cracked apart and, while some sections sank into Ganymede's interior, others remained floating on a sea of slushy ice. The crust eventually froze over once again, forming the lighter-coloured boundaries between the ancient crust segments, but a mantle of moving ice remained underneath, pushing the crust around.

Callisto, the most distant of the four great moons, is the only one that was never heated throughout – either internally or externally. It is a planetary glitter ball, with a dark surface studded with dazzlingly bright craters. Callisto's greatest mystery is the evidence for a worldwide saltwater sea several miles deep below its crust. Astronomers are still trying to understand how such a sea could have formed.

Galileo's great moons This composite photograph shows the Galilean moons in their order from Jupiter (from left to right): Io, Europa, Ganymede and Callisto.

See also: The home system 32 · Genesis 34 · Jupiter – king of the planets 64 · The Copernican revolution 192

Watching the Galilean moons

IO'S SURFACE

EUROPA'S SURFACE

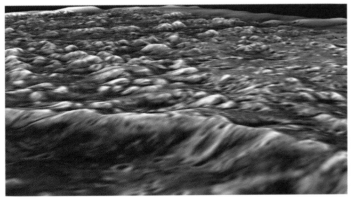

GANYMEDE'S SURFACE

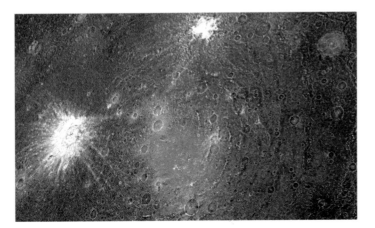

CALLISTO'S SURFACE

Callisto Europa Jupiter Io Ganymede

Callisto Io Jupiter Europa

Ganymede Callisto Jupiter Io Europa

Waltz of the moons The Galilean moons are easily seen with binoculars or a small telescope - if it were not for the brilliant glare of Jupiter itself, they would be visible to the naked eye. Lined up in a neat row on either side of Jupiter, they present a fascinating and ever-changing spectacle.

Shadow dancing The satellites have an equatorial orbit, often passing in front of, or behind, the planet in events called transits and occultations. When a moon transits the face of Jupiter, it casts a shadow onto the clouds below, which can be seen with a telescope.

STAR STORIES

Jupiter's Attendants

Jupiter's satellites have all been named after the various attendants and lovers of Zeus - Jupiter's counterpart in Greek legend—after a suggestion by Johannes Kepler, the German astronomer and contemporary of Galileo. Io was a priestess of Hera, Zeus' own wife, with whom the King fell in love. To hide her from Hera, he disguised her as a heifer. Europa was the daughter of the King of Phoenicia, whom Zeus carried off while disguised as a bull. She gave birth to a son, Minos, who became king of Crete, famous for its cult of bull worship. Ganymede was a handsome youth that the King of the Gods carried off to be cupbearer at the feasts of Olympus. Callisto was a water nymph and follower of the hunter-goddess Artemis, who killed her when she let herself be seduced by Zeus.

68 Saturn and its moons

The ringed world of Saturn lies sixth from the Sun, and was the most distant world observed before the telescope was invented. It is a gas giant, slightly smaller than Jupiter, with cloud belts of muted cream rather than the bright colours of its neighbour.

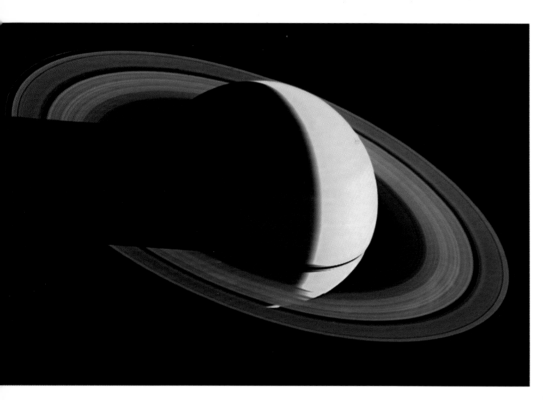

The ringed planet The most impressive feature of Saturn is its system of bright rings, extending thousands of miles from the planet itself and divided by dark gaps. The main ring plane is remarkably narrow - only several kilometres thick at most.

PLANET PROFILE

Visual magnitude (avg.)	0.7
Distance from Sun (avg.)	1,431 million km
	889.8 million miles
	9.57 AU
Length of year	29.37 Earth years
Rotation period	10.65 Earth hours
Equatorial diameter	120,536 km
	74,898 miles
Mass (Earth = 1)	95.1
Gravity at equator (Earth = 1)	1.1
Cloud-top temperature	-140 °C
	-220 °F

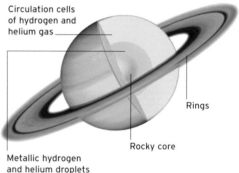

Circulation cells of hydrogen and helium gas

Rings

Rocky core

Metallic hydrogen and helium droplets

Saturn formed about 4.5 billion years ago in the same part of the Solar System as Jupiter. Although both planets have a similar mix of gases, their appearances differ, owing to their distances from the Sun and the contrasting temperatures of their cloud-tops. Farther from the Sun than Jupiter, Saturn has the colder atmosphere. Ammonia gas in the upper atmosphere condenses into ice crystals, forming a thin cloud layer and veiling what lies beneath. Peering through this shroud would reveal a world just as turbulent and colourful as Jupiter.

Saturn's internal structure is also probably quite similar to Jupiter's, with a hot rocky core surrounded by layers of metallic hydrogen, and liquid and gaseous hydrogen and helium. Curiously, although Saturn's mass is just 30 per cent of Jupiter's, it still occupies 60 per cent of Jupiter's volume – giving it a density less than water. This situation arises because Saturn's smaller mass gives it lower gravity, allowing its outer layers to expand more. For gas giants such as Jupiter and Saturn, a greater mass tends to increase the planet's density far more than its volume.

Saturn is orbited by a huge variety of moons – at least 30 at the time of writing, with new ones added every couple of years. Most are quite small, but eight are more substantial worlds– Mimas, Enceladus, Tethys, Dione, Rhea, Titan, Iapetu, and Phoebe.

Most of these moons seem to be made of rock and frozen water. Their surfaces range from heavily cratered to newly resurfaced and, as at Jupiter, the amount of resurfacing activity is related to the size of the moon and its proximity to Saturn. A moon's size affects the amount of internal heat produced during its formation, whereas its closeness to Saturn affects the degree of gravitational tidal heating it receives.

See also: The home system 32 · Genesis 34 · Jupiter - king of the planets 64 · Saturn's rings 70

Finding and observing Saturn

Saturn locater chart *Saturn is the faintest of the five naked-eye planets, yet it still appears as a bright star of around magnitude 0.7 in the night sky, and has a distinctive yellowish colour. It orbits the Sun once every 30 years and, because its orbit dwarfs Earth's own, it barely changes brightness between opposition and conjunction.*

The Saturn locater allows you to work out the approximate position of Saturn at any time up to 2030. The outer ring shows Saturn's position in the zodiac and also the tilt of Saturn's rings. The inner ring shows Earth's position throughout the year. Where the line from Earth to Saturn passes to the right of the Sun, the planet can be seen rising before the Sun in the predawn sky. Where it passes to the left, it sets after the Sun in the evening sky.

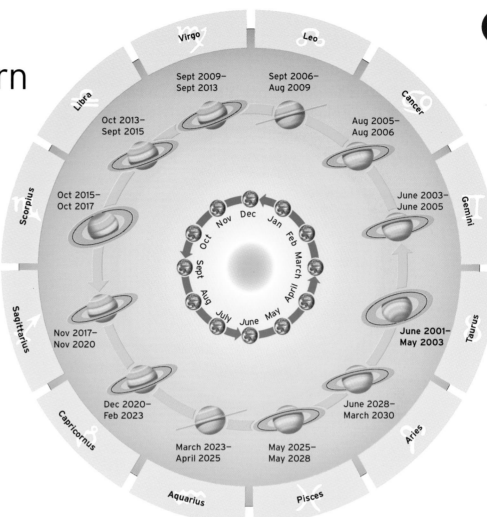

Coloured filters Subtle differences in Saturn's clouds can be viewed through coloured filters. Blue and yellow-green filters are most useful, helping to reveal and exaggerate the belts, and white spots that appear when a major storm forms in Saturn's atmosphere. A blue filter may also reveal a curious and unexplained effect where the rings appear much brighter to the west of the planet than to the east.

Mystery moon Saturn's largest satellite, Titan, is the only moon in the Solar System with a substantial atmosphere - it is smothered in a thick layer of orange haze. Studies of the moon's light have revealed that its atmosphere, like Earth's, is made up largely of nitrogen, and that it also has substantial amounts of methane - responsible for its colour. A spaceprobe is due to land on this mysterious world in 2004.

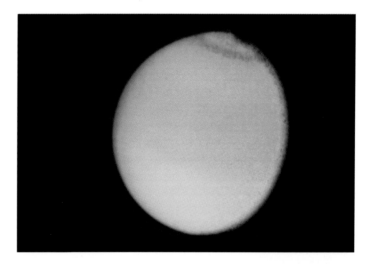

SATURN UNFILTERED

BLUE FILTER

YELLOW-GREEN FILTER

70 Saturn's rings

Saturn's most beautiful feature is its vast and brilliant rings. Although it is not the only planet in the Solar System to have rings, most are far less impressive. So what are the origins of these spectacular structures?

In 1850, physicist and mathematician James Clerk Maxwell came to the conclusion that Saturn's rings must be composed of countless individual objects, each in its own separate orbit around the planet. In the 1970s, the first spaceprobes to visit Saturn confirmed his theory, revealing that each major ring consists of thousands of ringlets.

Astronomers have since probed the individual ring particles and found that they are largely made of water ice – from house-sized pieces in the bright rings to tiny specks in the fainter ones. They also found that fine particles were gradually sifting out of the rings and falling into Saturn's atmosphere.

Thin, dark rings were found around Uranus in 1977 and Neptune, too, is now known to have a system of narrow rings. Even Jupiter has proved to have a previously unsuspected ringlet of dust surrounding it. So it seems that giant planets naturally form rings – but how?

It seems that rings form when an icy body breaks up close to a giant planet. The rings' progenitor may be a fragile icy moon, blasted apart in a collision, or, as was probably the case for Saturn, a passing comet pulled to its doom by the planet's gravity.

The break up scatters chunks of debris in wild orbits around the planet, each colliding with others, shattering and shifting orbit. Over a few thousand years, the debris is steered by collisions into circular orbits – where there is less chance of pieces hitting each other – in a flat plane above the planet's equator.

If this theory is correct, then bright rings like Saturn's must be reasonably fresh. Particles are still constantly colliding, grinding down into dust and drifting toward the planet. Over tens of millions of years – the blink of an eye in astronomical terms – a magnificent system like Saturn's will gradually decay into a ghost of its former self.

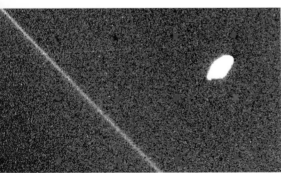

False appearances Although the rings appear solid from a distance, a close-up photograph reveals thousands of semitransparent ringlets.

Shepherd moons Some narrow ringlets are kept in place by tiny satellites just inside and outside their orbits. The material that forms the rings may actually be chipped off these moons by micrometeorites.

See also: The home system 32 · Genesis 34 · Asteroids 62 · Saturn & its moons 68 · Probes to the planets 202

Viewing Saturn's rings

Changing aspects *Saturn's equator is tilted at 26.7 degrees to its orbit, which means that the planet has seasons just like those on Earth. From our viewpoint, the tilt of the rings appears to change slowly over 30 years. Once every 15 years, the rings disappear almost completely as they appear edge-on to Earth. The next such events will happen in 2009 and 2025. After seven and a half more years, the rings appear wide open as we look across their plane either from above or below. As the orientation of the rings changes, so does the angle of the shadow they cast on the disc of the planet.*

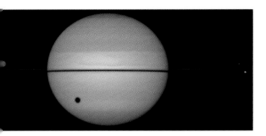

Disappearing rings Saturn's rings are only a few miles wide at their thickest, and they disappear completely through small telescopes when seen edge-on from Earth. This a good opportunity to see the fainter moons closer to the planet or to search for undiscovered satellites.

STAR STORIES

Saturn's Ears

Before 1655, when Christiaan Huygens discovered the true nature of Saturn's rings, the planet's strange shape was the subject of much debate. Galileo Galilei believed it had two giant moons on either side of it. When the rings briefly disappeared, he changed his mind to suggest that the planet had giant handles, or ears, extending from it.

Saturn

D ring Almost invisible from Earth, the faint and semitransparent D ring extends down toward the planet.

C ring The crepe ring is the faintest ring normally visible from Earth.

B ring This bright ring forms the inner segment of the rings as seen through a small telescope.

Cassini Division This gap between the A and B rings is not entirely empty – it contains four faint ringlets.

A ring The outer bright ring is divided in two by the Encke Division.

F ring This narrow ring runs around the outside of the A ring.

G ring The gravity of two shepherd moons, Prometheus and Pandora, keeps this narrow ring in line.

Ring system of Saturn The visible rings stretch to 172,000 km (107,000 miles) from the centre of Saturn – nearly three times the planet's radius.

E ring An outer halo of tiny particles stretching 480,000 km (300,000 miles) from Saturn forms the E ring.

The divisions in the rings The easiest features to detect from Earth are the A, B and C rings, and the Cassini and Encke Divisions (named after their discoverers). Such divisions are regions where the orbit of any lump of debris would regularly bring it into line with one of the major satellites. This sort of orbit is unstable, as the moon's insistent gravity tends to pull objects out of their circular orbits into paths where they may collide with each other or be thrown out of the rings altogether. The ring gaps are a small-scale replica of the Kirkwood gaps in the asteroid belt.

72 Uranus - the turquoise gas giant

For two centuries, little was known about Uranus - the first planet to be discovered by telescope - except that it had several moons and orbited on its side. Only in the Space Age did this giant's most surprising secrets come to light.

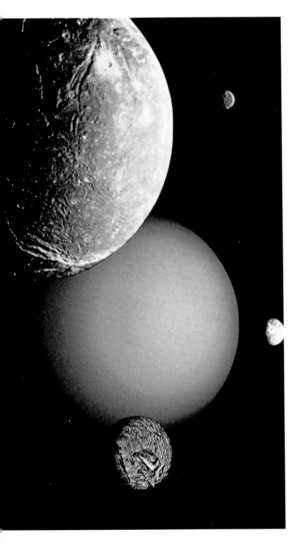

Uranus' moons Uranus has a vast family of moons - at least 18 - although only five were large enough to observe from Earth before the Voyager flyby.

Uranus orbits twice as far from the Sun as Saturn, and is roughly half its size. Its distinctive turquoise colour arises from a small percentage of methane gas in its atmosphere, which absorbs red light and reflects other colours. Even close up, Uranus seems mysterious – when Voyager 2 encountered the planet in 1986, it found a featureless world floating like a cue ball in space.

Uranus' strange lack of weather may derive from its seasons – the planet's poles only see sunrise once every 84 Earth years. In theory, this should mean that one pole is always in long-term deep freeze; however, the cloud-top temperature is strikingly even across the planet. This suggests that strong winds blow from warmer regions to colder ones, and these could prevent major weather systems developing.

Like all the gas giants, Uranus' outer layers are made mostly of the light gases hydrogen and helium. Deeper down, the planet contains large amounts of ices – chemicals such as ammonia and methane that evaporate into gas closer to the Sun. The high pressure inside Uranus turns these ices into an electricity-conducting slush, which swirls around, generating a magnetic field. Because Uranus' magnetism is not generated in its rocky core, the field is very strange – pointing sharply away from the planet's axis of rotation and not even passing through its centre. Astronomers think that this and the planet's 98 degree axial tilt may have resulted from a collision with another planet early in its history.

Uranus is surrounded by a system of thin rings that are very different from those around Saturn. Made up of methane ice rather than highly reflective water ice, they contain far less material than those of Saturn. Farther out, Uranus has a vast array of moons, made from a mixture of rock and ice, and showing various signs of past activity. The larger moons seem to have melted on the inside, then expanded as they froze, splitting their crusts apart. The most spectacular Uranian moon – and the closest major satellite to Uranus – is tiny Miranda, just 480 km (300 miles) wide. Miranda shows a bewildering range of terrains on its surface, bounded by sharp divisions and 3-mile- (5-km-) high cliffs. Astronomers think it was once melted and dramatically reshaped by tidal heating.

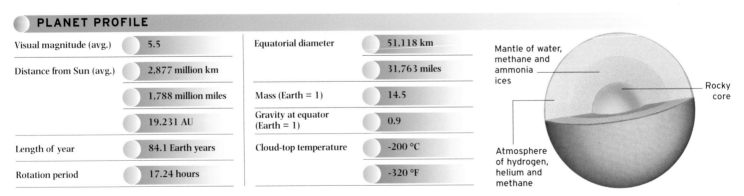

PLANET PROFILE

Visual magnitude (avg.)	5.5	Equatorial diameter	51,118 km	Mantle of water, methane and ammonia ices	
Distance from Sun (avg.)	2,877 million km		31,763 miles		Rocky core
	1,788 million miles	Mass (Earth = 1)	14.5		
	19.231 AU	Gravity at equator (Earth = 1)	0.9		
Length of year	84.1 Earth years	Cloud-top temperature	-200 °C	Atmosphere of hydrogen, helium and methane	
Rotation period	17.24 hours		-320 °F		

See also: The home system 32 · Genesis 34 · Saturn's rings 70 · Observatories & telescopes 180

Finding and observing Uranus

Uranus locater map *Uranus takes 84 years to circle the sky, spending several years in each constellation of the zodiac. Between 2001 and 2010, it is moving through Capricornus and into Aquarius. Its path against the background stars of this region is shown here. These constellations lie behind the Sun between December and March each year and, at an average brightness of magnitude 5.5, Uranus is only visible against a dark sky. The best time for viewing is between April and October. Early in this period, the constellations appear in the east before dawn, rising gradually earlier as the months pass. By October, they have become evening constellations, setting soon after the Sun.*

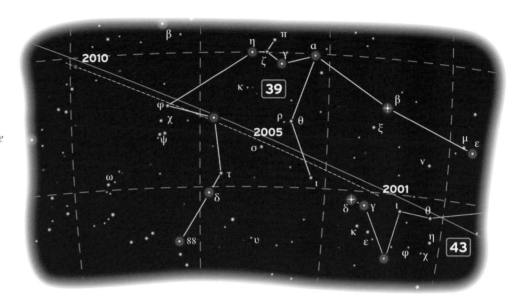

Uranus by night Through binoculars or an amateur telescope, Uranus appears as nothing more than a green star of magnitude 5.5. The best way to distinguish it from the background stars is to use a detailed locater chart compiled for the night you are observing. In theory, the planet is also visible to the naked eye under ideal observing conditions.

Sketching Uranus Earth-based telescopes reveal little of Uranus except its colour, but with a medium-sized instrument you can watch the dance of the planet's major moons. Titania, Oberon, Umbriel and Ariel are all 15th magnitude, and Miranda is magnitude 16.3. Because the moons orbit at right angles to the ecliptic, their paths reveal that Uranus itself is tipped over.

STAR STORIES

Enthusiast Discovers Uranus

Uranus was discovered by Sir William Herschel, a German-born musician who had come to England for his work. His passion for astronomy led him to build the finest telescopes of the day, which he used to chart the skies from his home in the town of Bath. His sister Caroline was also a keen researcher, assisting him in his work, and was one of the first women to make a significant contribution to astronomy.

74 Neptune - a dark and stormy world

Neptune, the eighth planet from the Sun and the last gas giant, is similar in size and colour to Uranus. A spaceprobe flyby in 1989, and subsequent observations with the Hubble Space Telescope, have revealed a world of stormy beauty and surprising activity.

Neptune is a gas giant about four times the diameter of Earth. Its atmosphere is made of light hydrogen and helium, with about 3 per cent methane, which absorbs red light, making the planet blue-green. Outer layers of atmosphere surround a slushy mantle of molten ices and the planet's rocky core.

In contrast to Uranus, Neptune has huge dark storms in its atmosphere, distinct bands around its equator, and high-altitude white clouds blown by 2,000-km/h (1,250-mph) winds – the most powerful in the Solar System.

Whereas weather on the rocky inner planets is driven by heat from the Sun, the immense gas worlds generate heat from inside – through their own steady contraction – which drives their weather. Neptune generates twice as much energy as it emits, explaining its activity. In the case of Uranus this internal power source has been switched off – possibly because of the planet's tilted axis.

Like all the gas giants, Neptune has a system of rings, more like the dark narrow rings of Uranus than the spectacular system around Saturn. Strangely, the rings are thicker on one side of the planet than the other.

While the inner gas giants have huge families of satellites, Neptune has just a handful. By far the largest is Triton, a moon bigger than our own, covered in a mixture of nitrogen, methane and carbon monoxide ices. This world's terrain looks as though it has been melted in the recent past. Most intriguing of all, Triton orbits Neptune in a circular orbit, but the wrong way around the planet.

Scientists think that Triton is a gate-crasher in the Neptune system which strayed into the giant's gravitational reach, and was then pulled into a circular orbit around the planet. As Triton spiralled inwards, it scattered most of Neptune's original satellites, leaving just two significant survivors – Proteus and Nereid. The extreme gravitational forces may also have heated Triton up, melting its surface.

Neptune, seen from orbit above Triton The planet has slightly more methane in its atmosphere than Uranus and hence is somewhat bluer.

PLANET PROFILE

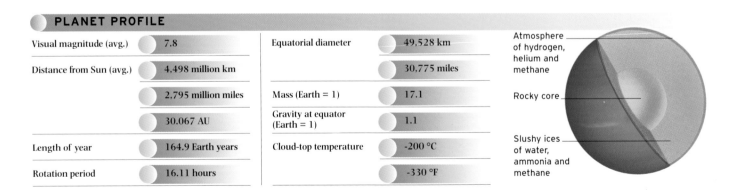

Visual magnitude (avg.)	7.8	Equatorial diameter	49,528 km	Atmosphere of hydrogen, helium and methane
Distance from Sun (avg.)	4,498 million km		30,775 miles	
	2,795 million miles	Mass (Earth = 1)	17.1	Rocky core
	30.067 AU	Gravity at equator (Earth = 1)	1.1	
Length of year	164.9 Earth years	Cloud-top temperature	-200 °C	Slushy ices of water, ammonia and methane
Rotation period	16.11 hours		-330 °F	

See also: The home system 32 · Genesis 34 · Uranus–the turquoise gas giant 72 · Probes to the planets 202

Finding and observing Neptune

Neptune locater map *Taking 164 years to orbit the Sun, Neptune crawls across the sky at a snail's pace, spending more than a decade in each constellation of the zodiac. Between 2001 and 2010, it moves from Capricornus into Aquarius. This locater map shows Neptune's path against the background stars of this region but, at magnitude 7.8, Neptune is only just visible through binoculars or a small telescope. Capricornus and Aquarius lie close to or behind the Sun between December and March each year, so viewing is best between April and October. Early in this period, the constellations appear to the east in the predawn sky, rising gradually earlier as the months go by. By October, they become evening constellations and set shortly after the Sun.*

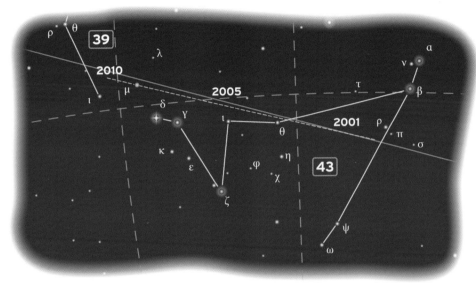

Neptune astrophoto Through a telescope, Neptune shows up as a faint green star with no distinguishable features. Its huge satellite, Triton, is brighter than any of Uranus' satellites but, shining at average magnitude 13.6, it is only visible through a large telescope.

STAR STORIES

Storms Over Neptune's Discovery

The story of Neptune's discovery is a controversial one. The British mathematician John Couch Adams and the French scientist Urbain Joseph Leverrier independently predicted its existence in the mid-19th century, based on irregular variations in the orbit of Uranus. Both had concluded – following formulas devised by Newton – that the irregularities in Uranus's orbit must be caused by the gravitational pull of another planet beyond its orbit. By 1843, Adams had calculated its position, mass, and orbit. Recently graduated, however, he was considered too young and inexperienced to be taken seriously, and could not convince professional astronomers to take notice. Leverrier was more fortunate when, three years later, he sent his calculations to Johann Gottfried Galle, then assistant director at the Berlin Observatory, where new star charts had recently been completed. Galle set a search for the planet in motion and became the first astronomer to observe Neptune when he discovered it, exactly where Adams and Leverrier had predicted it would be.

Disappearing spots Neptune can be seen from Earth with binoculars if its exact position is known, although it appears simply like a star. Planetary scientists using the Hubble Space Telescope (as seen here) have detected cloud features in Neptune's atmosphere that come and go over the months and years.

76 Pluto and the Kuiper Belt

Pluto, the most distant known planet, is a tiny world in darkness. Astronomers now argue as to whether it qualifies as a planet or is, in fact, just the largest of many icy asteroids that make up the Kuiper Belt beyond the orbit of Neptune.

Astronomers' best ideas about the nature of Pluto come from studies of Neptune's satellite Triton. All the evidence suggests that Triton is an icy world like Pluto that happened to stray into orbit around the gas giant. Comparison of light from both suggests the two worlds have similar surface chemical compositions – carbon monoxide, nitrogen and water ices.

Mapping the remote world Pluto remains unvisited by spaceprobes, but in 1996 the HST mapped its surface, identifying 12 major regions.

Pluto's orbit is highly eccentric – the wildest of any planet in the Solar System. At closest approach to the Sun, Pluto comes inside the orbit of Neptune and, when farthest away, it lies at 1.6 times Neptune's distance. This sort of orbit is more typical of asteroids or comets.

From Pluto, the Sun appears as a starlike point of light – although still 250 times brighter than the Full Moon. The tiny world is not alone in its long orbit – it has a moon, Charon, discovered in 1978. Charon is huge for a satellite– about half the size of Pluto – and orbits just 19,400 km (12,000 miles) away from its parent planet. Charon's presence has allowed astronomers to discover more about the system. For example, they now know that Pluto orbits on its side and that its rotation period and Charon's are the same as Charon's orbit about Pluto, so each is locked with one face permanently toward the other. This unique arrangement made it possible to map the surfaces of the two worlds during the late 1980s, when their positions were such that they eclipsed each other.

Although Pluto is officially a planet, hundreds of other rock-ice bodies are now known to orbit in the region beyond Neptune, some approaching the size of Charon. These Kuiper Belt objects are icy planetesimals left over from the genesis of the Solar System. Smaller chunks of ice orbiting among them

form short-period comets when they fall in toward the Sun. The Kuiper Belt might well contain other worlds larger than Pluto and, if one of these is discovered, it would probably force the reclassification of our tiny ninth planet.

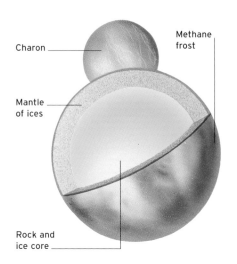

Charon

Methane frost

Mantle of ices

Rock and ice core

Pluto's moon Charon is fully half Pluto's size, making the pair a true double planet system. Its surface appears greyer in colour and contains more water ice than Pluto. The pair may have formed when another body collided with the early Pluto, blasting material into orbit to form Charon. In the process, Pluto's rotation axis could have been tilted.

See also: The home system 32 · Genesis 34 · Asteroids 62 · Neptune - a dark and stormy world 74 · Comets 78

The outer worlds

Planet of ice *Shining in the depths of the Solar System at a feeble magnitude 14.0, Pluto is invisible in most amateurs' instruments, and looks starlike even in the most powerful Earth-based telescopes. In the spring of 1996 the Hubble Space Telescope captured images of the planet that are the first to show Pluto's surface in any reasonable detail. Because of the tilt in its orbit, Pluto passes through constellations away from the zodiac. The locater map shows Pluto's path through Ophiuchus and Sagittarius between 2001 and 2015, allowing you at least to see Pluto's location, if not the tiny planet itself. For a more detailed map, consult the Internet or astronomical software.*

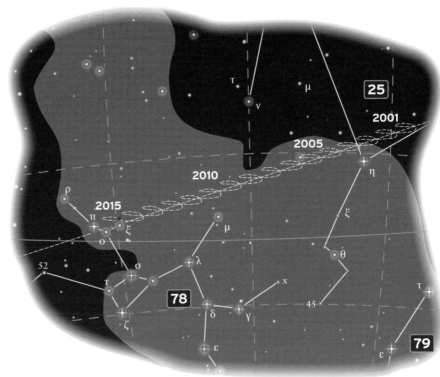

Kuiper Belt objects (below) Modern astronomers use the same basic technique that revealed Pluto to identify icy objects in the outer Kuiper Belt. These long-exposure photographs, taken three hours apart, show the same star field. The Kuiper Belt object, indicated by the circle, moves slightly between each. The brighter streak moving across the field between the two exposures is a much faster-moving and closer asteroid.

Pluto discovered (above) Mathematicians had predicted the existence of a ninth planet from wobbles in the orbits of Uranus and Neptune. In 1930 young astronomer Clyde Tombaugh discovered it using a telescope specially built for the job. He began his search near its calculated position. By taking two photographs of a star field several nights apart, and comparing them using an instrument called a blink comparator, he was able to identify any objects that were moving. He was very lucky – recalculations of Neptune's orbit have since proved that there were no unexplained wobbles, and Pluto just happened to be close to the predicted position at the time.

78 Comets

Beyond the orbit of Neptune lie the realms of comets – the Kuiper Belt where Pluto and other small worlds orbit, and the far more distant Oort Cloud, a spherical shell of icy debris left over from the Solar System's formation.

Most of the time, comets remain in deep-frozen "sleep" in outer space, but occasionally they are dislodged from their orbits and fall toward the inner Solar System. Comets arriving from the depths of the Solar System are still in the process of "waking up" and can be disappointingly unimpressive. The solid part of a comet, called the nucleus, is a lump of dirty ice, often just a few miles across, coated with a thin crust of dark, organic – carbon-based – chemicals that reflect sunlight poorly. As a result, comet nuclei are some of the darkest objects in the Solar System. But as the comet falls in past the orbit of Jupiter, something begins to happen.

Because its dark surface is a poor reflector of sunlight, the nucleus absorbs large amounts of light and heat. The ice under the dirty top layer begins to melt and instantly vaporizes, building up pressure beneath the crust. When this becomes too great, streamers of gas burst out into space. Ice exposed to sunlight quickly turns into gas, and soon the comet is surrounded by an expanding halo of gas – the coma. As the comet gets closer to the Sun, its coma increases in size and can develop into one or more tails stretching across millions of kilometres of space.

A comet's light comes from two sources—simple reflection of sunlight and emission of energy as the solar wind of particles from the Sun slams into gas particles in the coma. The gases of the coma are ionized by sunlight, and they are blown away from the Sun, forming a straight tail that glows bluish. Meanwhile, heavier dust particles lag behind the nucleus and tend to form a curving tail that shines yellowish-white as it reflects sunlight.

The behaviour of comets is highly unpredictable. Comets large enough to be detected in the outer Solar System may prove to have such thick, insulating crusts that they never develop, whereas smaller comets may be as insubstantial as a snowball and melt away to fragments as they come closer to the Sun. Even short-period comets that return to the inner Solar System every few years develop only a small coma – their frequent passages past the Sun have driven off most of their volatile ices.

Comet Hyakutake Several new comets sweep through the inner Solar System each year, but bright naked-eye comets only appear every couple of decades on average.

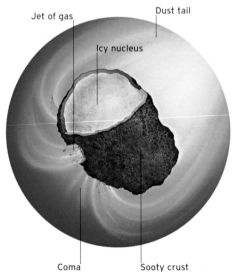

Dirty snowballs The core of a comet is surrounded by layers of ice and a black, sootlike crust of dust. Comets become visible as they heat up, releasing jets of trapped gas from beneath the crust. This stream of particles glows vividly as it gains energy from the solar wind.

Jet of gas

Dust tail

Icy nucleus

Coma

Sooty crust

See also: The home system 32 · Genesis 34 · Meteors 48 · Asteroids 62 · Pluto & the Kuiper Belt 76

Comet spotting

Spectacular visitors *Even faint comets are fairly easy to locate with the aid of an accurate locator chart and a pair of binoculars – their coma makes them look distinctively smudged and much larger than a pointlike star. The comet moves from night to night against the background of stars, picking up speed and growing its tail as it falls toward the Sun. The length of a comet's tail depends on the angle at which we see it – spectacular comets have tails covering as much as 30 degrees of the sky (60 times the width of the Full Moon) but, if we see the comet almost head-on, the tail can be drastically foreshortened. The outer edges of the tail are also very faint, so the full glory of a comet can only be appreciated from a dark site on a moonless night.*

Comet hunting Professional observatories rarely have time to devote to scanning the skies in search of new objects, so comets are often discovered by amateur astronomers using simple equipment. Because comets are at their brightest when they are closest to the Sun, it is best to search for them in the sky directly after sunset – although never when the Sun is still above the horizon. A basic technique is to sweep the horizon above the point at which the Sun set, gradually moving higher into the sky. Sun-grazing comets are the most elusive, passing close to the Sun and sometimes even plunging into it.

Comet orbits A comet's tail always points away from the Sun, whichever direction it is travelling in, so a comet travels headfirst on its approach to the Sun, and tailfirst as it recedes. The Sun's gravity usually causes the dust trail to curve in the direction of the comet's motion. The elliptical orbit of a comet also causes it to accelerate as it approaches the Sun. The nearer a comet goes to the Sun, the faster it moves – comets that pass close to Earth can travel across several degrees of sky each night.

Bearers of Ill Fortune

Throughout history, comets have been seen as portents of doom or signs of turbulent times to come. When Halley's Comet appeared in the skies of Europe in 1066, King Harold of England was warned that it foretold his downfall, and William of Normandy saw it as a good omen. A few months later, England was William's. Plagues and famines were thought to strike while comets hung in the sky, and, even as late as the nineteenth century, astronomers thought that the tail of a comet might contain vapours harmful to life on Earth.

No one quite knows how comets became ill omens – some astronomers have suggested that a giant comet might have fallen into a short-period orbit about the Sun in late prehistoric times. Its passages past Earth would have been accompanied by dangerous showers of meteorites. If this is true, then such a traumatic event could well have lingered in myth and legend for thousands of years after.

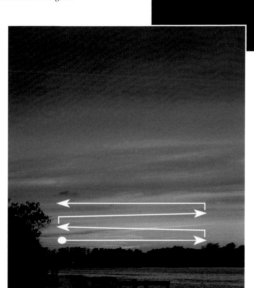

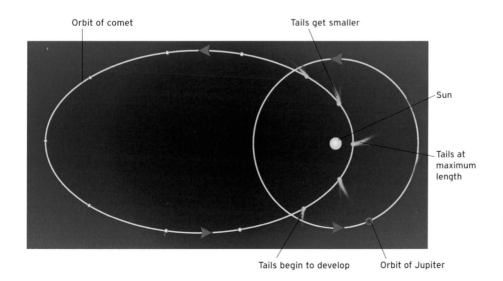

Orbit of comet

Tails get smaller

Sun

Tails at maximum length

Tails begin to develop

Orbit of Jupiter

The gulf of night An imaginary vista across the plane of our galaxy encompasses many stages in the lives of stars. Nearby, a black hole pulls material from a blue-giant star. Farther away, new star clusters are being born in pink and blue emission nebulae, while dying stars cast off their outer layers in shell-like planetary nebulae. Beyond lies the plane of the Milky Way itself, obscured with dark gas and dust clouds and orbited by old, yellow stars in globular clusters.

Stars and the galaxy

The starry sky

On a clear, dark night the sky appears be filled with millions of stars but, in fact, just a few thousand are visible to the naked eye. Unlike the planets, stars seem to follow an unvarying course and remain remote, even when viewed through a telescope.

By simply looking up at the night sky, two features are immediately obvious. The stars have different brightnesses and colours, and their distribution across the sky is uneven – some areas are thickly strewn with stars, whereas other areas are quite barren. A prolonged observation of the night sky reveals that the stars are more distant than anything in the Solar System because the Moon and planets move faster, on occasion passing in front of, or occulting, them.

Stars shine by their own light, so their different apparent brightnesses reveal that they must either have different true brightnesses (luminosities), or be at different distances from Earth. In fact, both explanations are true. The distribution of stars also reveals a great deal about the Universe and helps solve a famous astronomical question, known as Olbers' Paradox, which asks why the sky is dark at night.

If stars were randomly scattered through an infinite Universe then, whichever direction we looked, we should eventually see a star. Even though the more distant stars would be fainter, they should be more densely packed, and the end result would be a sky ablaze with starlight. The fact that much of the sky is dark allowed 19th-century astronomers to start reaching important conclusions about the Universe – either it is not infinite in size, and/or it has not existed forever, so the light from the most distant stars has not yet had time to reach us.

Humans, with their natural tendency to look for order in the Universe around them, have long ordered the stars into constellations. These groupings are a convenient way of mapping the night sky, but in reality they are just line-of-sight effects. The stars within them are rarely close to each other in space.

However, the tighter groupings of stars in close pairs and clusters is too numerous to be caused by simple line-of-sight effects. Binoculars reveal that the Milky Way – the long pale band stretching around the sky – is made up of countless close-packed stars, and is brighter due to our looking straight across a narrow plane of space where stars are concentrated.

All stars appear to be fixed in set positions as they circle the sky each day as Earth rotates. This, however, is an illusion caused by their immense distance from us. Measurements over the years have shown that most stars are moving on their own paths. To measure these proper motions, astronomers have to remove effects caused by the movements of Earth and the whole Solar System.

The night sky Billions of stars lie scattered across the plane of our galaxy. The concentration of stars in this narrow band creates a visibly brighter area in the night sky - the Milky Way.

See also: Star distances 84 · Star colours 88 · Plotting the lives of stars 90 · The Milky Way 134

Finding patterns in the night sky

34 URSA MAJOR *The Great Bear, Ursa Major, is probably the most familiar of all constellations in the northern sky. Its seven brightest stars are known by a variety of nicknames, most famously the Big Dipper and the Plough, but the constellation extends well beyond this central region, and is one of the largest in the sky. Ursa Major is an unusual constellation in that many of its stars are in the same region of space. Five of the Plough stars form part of the nearest star cluster to Earth, which is located about 80 light-years away and is known as the Ursa Major Moving Cluster.*

M81 One of the sky's finest spiral galaxies, M81 shines at magnitude 7.9 and is easily visible with binoculars as an elliptical smudge of light. A small telescope will show the concentration of stars at its centre more clearly, but it takes a large telescope to discern the spiral arms. It lies 10 million light-years away.

M101 Shining at magnitude 8.2, M101 is a large spiral galaxy, about twice as far away as M81. This galaxy displays its spiral arms directly to us but, as the light is spread over an area the size of the Full Moon, it appears dim. As a result, small telescopes and binoculars only reveal the light from its central regions. To see its structure clearly, long-exposure photographs or a larger telescope are needed.

ζ UMa The middle star of the Dipper's handle, Mizar is a famous double star of magnitude 2.3. Binoculars or sharp eyes can spot its nearby magnitude 4.0 companion Alcor. However, when you turn even a small telescope on Mizar, you discover that it is double. All of these stars are binary stars, whose companion stars are too close to separate even with a powerful telescope.

ξ UMa Xi Ursae Majoris is a double star, whose members of magnitudes of 4.3 and 4.8, can be separated through a small telescope. Both are yellow, and each is itself a double star. The stars orbit each other every 60 years and were closest together in 1992.

STAR STORIES

Bears in Space

The constellation we know today as Ursa Major was first identified with a bear in Ancient Greece, but North American Indians also saw the same animal in the same pattern. In Greek myth, Ursa Major represents the nymph Callisto, turned into a bear by her lover Zeus, the King of the Gods. Ursa Minor represents Callisto's son, Arcas, who mistakenly shot his mother and was then turned into a bear himself.

α UMa **β UMa** At the centre of Ursa Major lies Alpha Ursae Majoris, also known as Dubhe, a double star that shines at magnitude 1.8. The main star is a yellow giant located 124 light-years away. Its companion can only be separated by a large telescope. Below Dubhe lies Beta Ursae Majoris, also known as Merak, a white star of magnitude 2.4, which is situated 79 light-years from Earth. Dubhe and Merak together form the Pointers – a line through them points to Polaris, the northern pole star.

84 Star distances

Once they know the distance to a star, astronomers can work out information such as its true brightness and the path it takes through space. But how do they measure the distance of objects so far away that their light takes many centuries to reach us?

The astronomical distance scale hangs together like a chain. Astronomers have obtained very accurate distances to the nearest stars using basic principles of geometry. From these they have pieced together rules that allow them to work out the distances of more distant stars without measuring them directly. Each link in the chain takes distances thousands of times farther out into the universe – but the accuracy of the measurements for these more distant objects is only as good as the reasoning that supports them.

The first link in the chain relies on the principle of parallax – that nearby objects appear to change their position against a more distant background when our point of view changes. This effect is easily seen by extending a finger at arm's length and viewing it first through one eye, then the other. In theory, you could work out the distance to your finger by measuring the separation of your eyes – the baseline – and the position of your finger from each point of view.

The parallax effect diminishes for more distant objects but increases with the separation of the two viewing points. Fortunately, our planet provides us with a very long ready-made baseline – the opposite sides of Earth's orbit. Although this distance (300 million km or 186 million miles) is tiny on an astronomical scale (equivalent to 17 light-minutes), it is large enough to affect our point of view for stars in our local neighbourhood.

The first star to have its distance successfully measured using parallax was 61 Cygni, a faint star 10.3 light-years away in the constellation Cygnus. German astronomer Friedrich Wilhelm Bessel pioneered the method in the 1830s. It involved measuring changes in position for less than one second of arc. In order to get an accurate measurement, he had to work out how to compensate for the star's own movement through space, the bending of its light rays by Earth's atmosphere, and the imperfections of his own instruments.

Parallax measurement was soon used to work out the distances to hundreds of the closest stars, and it is still the most precise way of measuring interstellar distances. In fact, professional astronomers often use the parallax second (parsec), the distance at which an object would have a parallax of exactly one second of arc, as a measure of distance instead of the light-year (one parsec is equivalent to 3.26 light-years.)

The amount of light reaching Earth from a star diminishes with distance. So, by combining parallax with measurements of apparent magnitudes, astronomers could for the first time work out the true luminosities of stars. This revealed that they range from dim dwarfs to brilliant giants, with our Sun roughly in the middle.

The parallax method As Earth orbits the Sun, nearby stars appear to move from side to side against a backdrop of more distant stars. The angle through which any star moves over a period of six months is called its parallax. The star's distance can then be calculated using simple geometry - the larger the parallax, the closer the star.

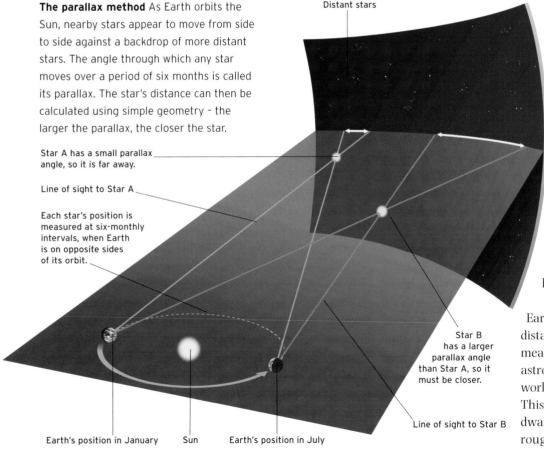

Distant stars

Star A has a small parallax angle, so it is far away.

Line of sight to Star A

Each star's position is measured at six-monthly intervals, when Earth is on opposite sides of its orbit.

Star B has a larger parallax angle than Star A, so it must be closer.

Line of sight to Star B

Earth's position in January   Sun   Earth's position in July

See also: Nearby stars 86 · Plotting the lives of stars 90 · Distances to galaxies 146 · How astronomers work 178

Racing stars

Barnard's Star This faint, magnitude 9.5 red dwarf has the fastest movement across the sky of any star – 10 arcseconds per year, nearly 20 times greater than its parallax. Rapid motion is a much clearer indication that a star is near to us than parallax. Barnard's Star is, in fact, only 6.0 light-years away.

25 31 OPHIUCHUS & SERPENS *Known as the Serpent Bearer, Ophiuchus is a large and relatively faint constellation on the celestial equator. Serpens represents the snake he is carrying and is the only constellation split across two areas of sky – Serpens Caput, the head, and Serpens Cauda, the tail. Ophiuchus is regarded as the unacknowledged 13th constellation of the zodiac. It cuts across the ecliptic and most of the planets pass through it at various times. The southern end dips into the star fields of the Milky Way but most of Ophiuchus and Serpens lie just to the north of the galactic plane. This makes them especially rich in globular star clusters. Ophiuchus also contains Barnard's Star, the second-closest star to the Sun.*

RS Oph The faint star RS Ophiuchi normally shines at around magnitude 11.5, beyond the reach of small telescopes. However, it occasionally flares up to naked-eye brightness (around magnitude 3.5) in a bright explosion that lasts a few weeks. Astronomers call such a star a recurrent nova.

M10 Globular cluster M10, 14,000 light-years away, is one of the finest globulars visible in the northern hemisphere. Shining at magnitude 6.6, it appears in binoculars or smaller telescopes as a fuzzy star half the size of the Full Moon. Medium-sized telescopes can easily resolve the stars around its edges.

70 Oph The double star 70 Ophiuchi consists of a yellow, magnitude 4.2 star orbiting with a red, magnitude 5.9 companion roughly 16 light-years from Earth. The stars orbit around each other once every 88 years. At their widest separation around 2010, they will be easily separable with small telescopes but, at their closest approach, larger instruments are needed.

M5 M5 is another spectacular globular cluster, visible to the naked eye on dark nights. It is 24,500 light-years away and shines at magnitude 5.6. A small telescope will show its bright central core surrounded by fainter outer regions, but larger instruments are needed to see the individual stars.

 STAR STORIES

The Serpent Bearer

Ophiuchus, an Arabic name, is often associated with Asclepius, the Greek god of healing. This seems to have been the first name for the constellation, as early as the fourth century B.C., but it was soon associated with several dragon-wrestling heroes, including Hercules. Asclepius was the surgeon aboard the Argo, the ship of Jason and the Argonauts. In ancient sculptures he is often represented with a staff with a serpent coiled around it.

ρ Oph With four components of magnitudes 5.0 to 7.3, Rho Ophiuchi is an easily resolved multiple star for small instruments. It is surrounded by IC 4604, a cloud of dust and gas the width of the Full Moon, only visible in long-exposure photographs.

Nearby stars

There are several known stars within ten light-years of our Solar System. The closest of these are the three members of the Alpha Centauri system, lying a mere 4.3 light-years from Earth. These near-neighbours allow us to observe stars close-up.

All stars are not the same – a fact that is clearly illustrated by the surprising variety in the Sun's neighbourhood. To make sense of the many different types of stars, astronomers have developed measuring techniques to collect as much information about their composition and movement as possible.

The nearest known star is an 11th magnitude dwarf star, Proxima Centauri. This faint star lies at the limit of visibility for small telescopes and was neglected for years until its distance was measured and it turned out to be 0.11 light-years closer to us than its bright near-neighbour in the sky, Alpha Centauri. The proximity of these two nearby stars suggests that they are probably held together by gravity, but this is hard to verify because Proxima's movement around the larger star is so slow – it is calculated that one orbit would take at least one million years. To complicate matters further, Alpha Centauri is a double star, consisting of a bright yellow star and a somewhat fainter orange companion, locked in an 80-year orbit around each other. In addition, Proxima Centauri is a variable star, frequently changing its brightness by as much as one order of magnitude. Compared with this complex system, our lone-Sun system is relatively mundane.

Farther afield, the range of different stars rapidly becomes bewildering. The next nearest star is Barnard's Star, a dim red star in Ophiuchus that shifts its position compared with its neighbours very noticeably. This proper motion reveals that the stars are not fixed relative to each other, but are constantly on the move. Beyond this lie another four red dwarfs, two of which form another binary system – further evidence that both dwarf stars and multiple star systems are very common in our galaxy.

At a distance of 8.6 light-years, Sirius is the nearest really bright star to us. Because of its proximity and the fact that it is around 20 times more brilliant than the Sun, it is little wonder that Sirius is the brightest star in the night sky, with a distinctive white colour.

Sirius also has a companion star, Sirius B, which despite being very faint and small is white in colour. However, its orbit around Sirius indicates that this tiny "white dwarf" is surprisingly dense.

Because astronomical instruments have only just reached the point where they can detect the smallest and faintest stars, many astronomers believe that one or more of these dim dwarf stars could lurk even closer to the Sun than Proxima Centauri. Just because we have not seen any stars closer than 4.24 light-years from the Sun – a distance of 40 trillion km (25 trillion miles) – does not mean that they are not there. Some astronomers believe that a dwarf star could even be locked into a distant orbit around the Sun.

Home on Alpha Centauri This artist's impression shows the Alpha Centauri system from a hypothetical planet in orbit around the nearest star, Proxima. As the red-dwarf sun rises, a moon and the bright stars Alpha Centauri A and B can be seen in the dawn sky.

See also: Star distances 84 · Star colours 88 · Red & brown dwarfs 102 · Stellar heavyweights 104 · White dwarfs 118

Home to our nearest neighbour

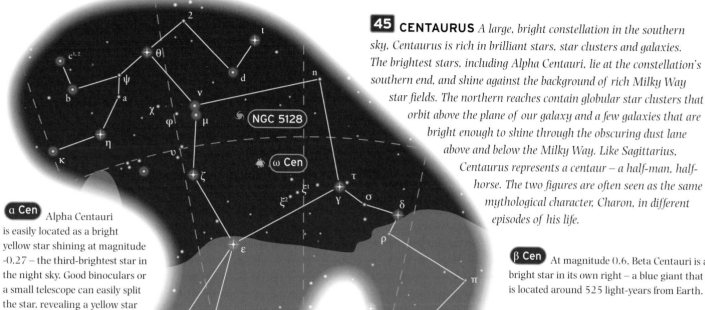

45 **CENTAURUS** *A large, bright constellation in the southern sky, Centaurus is rich in brilliant stars, star clusters and galaxies. The brightest stars, including Alpha Centauri, lie at the constellation's southern end, and shine against the background of rich Milky Way star fields. The northern reaches contain globular star clusters that orbit above the plane of our galaxy and a few galaxies that are bright enough to shine through the obscuring dust lane above and below the Milky Way. Like Sagittarius, Centaurus represents a centaur – a half-man, half-horse. The two figures are often seen as the same mythological character, Charon, in different episodes of his life.*

α Cen Alpha Centauri is easily located as a bright yellow star shining at magnitude -0.27 – the third-brightest star in the night sky. Good binoculars or a small telescope can easily split the star, revealing a yellow star of magnitude 0 and a more orange-coloured star of magnitude 1.3. Close to the southwest lies Proxima Centauri, magnitude 11.0 (shown by marker).

β Cen At magnitude 0.6, Beta Centauri is a bright star in its own right – a blue giant that is located around 525 light-years from Earth.

NGC 5128 Also known as Centaurus A after the strange source of radio waves contained within it, NGC 5128 is the brightest galaxy in Centaurus, lying 15 million light-years away and shining at magnitude 7.0. It appears as a smudge of light in binoculars, but a medium-sized telescope shows that it is actually a huge ball of stars, with a trail of dark dust across its centre.

ω Cen Shining at magnitude 3.7, Omega Centauri is easily mistaken for a star at first glance – it was even given a Greek letter designation. However, a closer look shows that it is a globular cluster containing millions of stars. Located 17,000 light-years away and measuring 650 light-years across, Omega Centauri is the largest known globular cluster and the most spectacular object in Centaurus.

OTHER NEARBY STARS

NAME	CONSTELLATION	DISTANCE	TYPE	MAGNITUDE
Barnard's Star	Ophiuchus	6.0	Red dwarf	9.5
Wolf 359	Leo	7.7	Red dwarf	13.5
BD +36°2147	Cygnus	8.2	Red dwarf	7.5
UV Ceti A/B	Cetus	8.4	Red dwarf	12.4/13.0
Sirius A/B	Canis Major	8.6	White main sequence/ white dwarf	-1.4/8.3
Ross 154	Sagittarius	9.4	Red dwarf	10.5
Ross 248	Andromeda	10.4	Red dwarf	12.3
Epsilon Eridani	Eridanus	10.5	Sunlike main sequence	3.7

88 Star colours

Once the distance to a star is known, astronomers can work out its true brightness and calculate the amount of energy it generates. By splitting the light from a star, astronomers can reveal its chemical composition and also obtain clues to the way it is moving through space.

Colours of the Universe This Hubble Space Telescope view of a dense star-cloud near the centre of the Milky Way reveals the huge range of colours that allow astronomers to probe the secrets of the stars.

Stellar spectra The light emitted by a star can be broken into a spectrum to analyse its components. Some frequencies of light are absorbed by the star's atmosphere and these show up as dark vertical bands.

Most stars give out a continuum of radiation of many frequencies. By using filters to block certain frequencies and select others, astronomers can measure the amount of energy a star is giving out within a certain frequency band and work out its surface temperature.

Another way to study the range of radiations produced by a star is to split its light into a spectrum, a rainbow of different frequencies. Stellar spectra contain dark absorption lines, which are caused by light being absorbed by atoms in a star's atmosphere. Because different atoms absorb different frequencies of light, astronomers can use these lines to analyse the make-up of a star's outer layers. From this data, they can work out what nuclear reactions are taking place deep within its core.

Spectra can also show how fast a star is moving toward or away from us. A phenomenon called Doppler shift changes the frequency, and therefore colour, of light from a moving source. This shift – toward blue when a star is approaching us and toward red when it is moving away – is most easily seen in the changing positions of spectral lines.

Stars have distinct colours created by the combination of different frequencies of light. White light is not a colour but is a balanced combination of all colours, from deep red to violet. Our Sun produces light that comes close to white, but an absence of blue light gives it a slight reddish-yellow bias. Sirius, the brightest star in the night sky, is almost pure white. Different colours of light are produced by stars of different temperatures – as the temperature rises the frequency of light emitted increases, shifting the colour of the star toward the blue end of the visible spectrum. With more heat, the radiation emitted shifts into the ultraviolet and beyond to X-rays and gamma rays, all of which are invisible to human eyes. Conversely, objects too cool to glow visibly emit invisible infrared radiation, or low-frequency radio waves.

See also: Star distances 84 · Plotting the lives of stars 90 · Stellar evolution 92 · The arrival of astrophysics 196

Colourful Canis Major

α CMa From Earth, Sirius is the brightest star in the sky due to its proximity – a mere 8.6 light-years away. Shining 23 times brighter than the Sun and 1.8 solar diameters wide, Sirius is in fact only a fairly average white star. It is orbited by a companion star, Sirius B, which is also white, but far fainter and smaller. Although this white dwarf shines at magnitude 8.7, it cannot be seen with binoculars or small telescopes, as it is lost in the dazzle of the other star.

42 CANIS MAJOR *The bright constellation Canis Major, the Great Dog, lies just south of the celestial equator and east of Orion. It represents one of Orion's two hunting dogs, following the celestial hunter across the sky. The Milky Way runs across the constellation, making it rich in star clusters, and it also shows a wide range of different star colours and brightnesses. The brightest star in the constellation, and indeed the entire sky, is Sirius, the Dog Star. At magnitude -1.4, only Venus, Jupiter, Mars and Saturn can outshine this brilliant white star.*

M41 Over 2,400 light-years away, M41 is a bright open cluster that is visible as a Moon-sized smudge of light with the naked eye. Binoculars or a small telescope reveal the individual stars. The brightest of these are orange giants of seventh magnitude, but the majority of the cluster's stars are bluish white.

μ CMa Mu Canis Majoris is a yellow giant star of magnitude 5.0, located 900 light-years from Earth. It has a blue-white companion that has a magnitude of 9.0.

β CMa Shining at magnitude 2.0, Beta Canis Majoris, or Mirzam – the Announcer – is a blue giant star 500 light-years from Earth.

δ CMa Also known as Wezen – the Weight – Delta Canis Majoris is a yellow star 1,800 light-years from Earth. It shines at magnitude 1.8 and is an extremely luminous supergiant star.

NGC 2362 Centred on the magnitude 4.4 star Tau Canis Majoris, NGC 2362 is another open cluster, located about 5,000 light-years away. Many of the cluster's 40 stars can be seen with binoculars or a small telescope. Most are bluish white in colour and are also highly luminous. These are some of the youngest stars in the sky, probably all less than a million years old and still contracting. Tau itself seems to be a cluster member, and is one of the most luminous stars known, more than 50,000 times as bright as the Sun.

NGC 2354 The open cluster NGC 2354 lies just to the southwest of NGC 2362. Although it contains around 60 stars, it is a far looser assembly than its neighbour, covering an area roughly the size of the Full Moon.

STAR STORIES

The Dogon Tribe and Sirius B

Could a tribe from Mali in Africa, have known about Sirius B before its discovery by Western astronomers? This was the claim by French anthropologist Marcel Griaule, who visited the Dogon in the 1930s. According to Griaule, the Dogon called Sirius "Sigu Tolo", and talked of another star, "To Polo", that came close to it from time to time. Griaule's findings were a source of mystery for many years until another team of anthropologists visited the Dogon and found no credible evidence of this ancient wisdom.

Plotting the lives of stars

What can a star's colour tell us about it? Is there a link between a star's colour and its luminosity? When astronomers first asked this apparently simple question, no one realized that its answer would ultimately reveal the secrets of stellar life cycles.

Astronomers classify the colour of a star by its spectral type, indicated by letters from O – the bluest and hottest – through B, A, F, G and K, to M – the reddest and coolest. O-type stars can have surface temperatures up to 50,000 °C (90,000 °F), while M stars may be as low as 3,000 °C (5,400 °F). Each letter is in turn subdivided from 0 – the hottest – to 9 – the coolest.

To understand the relationship between spectral type and luminosity – a star's true brightness – Danish astronomer Ejnar Hertzsprung and US astrophysicist Henry Norris Russell independently invented the Hertzsprung-Russell (H-R) diagram in 1912.

When the colours and luminosities of stars are plotted against each other on an H-R diagram, they fall into a distinct pattern. Red stars are always very bright or very faint, yellow stars such as the Sun are mostly middle-of-the-range, and rare blue stars are the brightest. Stars higher up the diagram are called giants or supergiants, and stars lower down are called dwarfs. Our Sun is a sub-dwarf – just large and bright enough not to be a real dwarf star.

Most stars lie on a diagonal strip called the main sequence, which joins dim red dwarfs to brilliant blue giants. Stars that belong in other areas of the diagram, such as white dwarfs and red giants, are so rare that we can assume that most stars spend most of their lives on the main sequence. Bright stars at the top of the main sequence are also rarer than the dwarfs at its bottom end. Since bright giants can be seen over very long distances, they must be truly rare to be outnumbered by the much fainter dwarfs that can only be seen if they lie close to Earth.

The Hertzsprung-Russell diagram allows astronomers to arrive at an estimate of a star's true brightness – and thus its distance – without the need for direct measurements. They can tell a distant red giant from a nearby red dwarf by differences in their dark spectral lines – indicating their different compositions – and use this information to place the star in its correct place on the diagram. As astronomers began to plot thousands of stars on the H-R diagram, instead of just the nearby ones known previously, they found that the life stories of the stars were laid out before them.

H-R diagram labels (Luminosity compared to Sun): 1,000,000 · 100,000 · 10,000 · 1,000 · 100 · 10 · 1 · 0.1 · 0.01 · 0.001 · 0.0001 · 0.00001

Stars plotted: ρ Leo · η Leo · 72 Leo · ε Leo · α Leo (Regulus) · R Leo · δ Leo · ζ Leo · γ1 Leo · φ Leo · Sirius · γ2 Leo · β Leo · 40 Leo · 10 · χ Leo · α Centauri · 93 Leo · Sun · 1 · Sirius B · Barnard's Star · Proxima Centauri · Wolf 359

Spectral type / Temperature: O5 · B0 28,000 °C · A0 10,200 °C · F0 7,700 °C · G0 6,300 °C · K0 5,200 °C · M0 3,800 °C · M8 2,700 °C

Size matters Blue stars higher up the H-R diagram are brighter, so they must contain more fuel, and be larger. Red stars must also be larger than blue ones of similar brightness – they have cooler surfaces, so their energy must be escaping through a larger surface area.

Hertzsprung-Russell diagram
The H-R diagram plots a star's spectral type (and temperature) horizontally and its luminosity vertically. When a selection of stars from Leo are plotted with some other well-known stars, most fit neatly onto the main sequence from dim red to bright blue.

See also: Star distances 84 · Stellar evolution 92 · Sunlike stars 98 · Red & brown dwarfs 102 · Stellar heavyweights 104

Stars and galaxies in Leo

γ Leo Gamma Leonis (Algieba) is a beautiful double star, made up of two yellow giant stars 170 light-years away, shining at magnitudes 2.0 and 3.2, and orbiting each other every 600 years.

21 LEO *Representing a lion, Leo is one of the few constellations that actually resembles the creature it is named after. The pattern is easy to spot and is best seen in both hemispheres around March. Leo is a large zodiac constellation, so the Sun and planets pass through it regularly. But the constellation lies well away from the Milky Way, so it is relatively empty of stars and nebulae. The stars of Leo cover many spectral types and luminosities, mostly falling on the main sequence. Distant galaxies, including M65, M66, M95, M96 and M105, can also be seen here.*

M65 M66 Messier objects 65 and 66 are two spiral galaxies of 9th magnitude. Both are about 20 million light-years from Earth, and are visible with binoculars on a dark night. Just north of them is an edge-on spiral galaxy, only visible with a telescope.

M105 M105, 1 degree north of M96, is an elliptical galaxy of magnitude 10.5. It is only visible with a medium-sized telescope, and appears as a small fuzzy ball, since it is over 25 million light-years away.

α Leo Alpha Leonis (Regulus) is a blue-white star 130 times as luminous as the Sun, and 78 light-years away. It is one star in a double system, with a much smaller and fainter companion that is visible in binoculars.

R Leo R Leonis is a variable red star 3,000 light-years away. It cycles between 6th and 10th magnitude in 312 days, and gets redder as it gets brighter.

M95 M96 Messier objects 95 and 96 are another pair of spiral galaxies, of magnitude 10.5 and 10.1. M95 is a face-on, barred-spiral galaxy that lies 22 million light-years from Earth.

✦ STAR STORIES

The Magical Lion

The stars of Leo have been seen as a lion for as long as people have watched the skies, and the ancient Greeks attached a legend to it over 2,000 years ago. According to Greek writers, Leo represents the Nemaean Lion, which the hero Hercules fought as the first of his 12 labours. The lion was created by the Moon goddess Selene and had a pelt impervious to weapons, so Hercules had to wrestle it bare-handed. Hercules later wore the animal's skin as armour and its head as a helmet.

92 Stellar evolution

The Hertzsprung-Russell (H-R) diagram plots all kinds of stars by their temperatures (colours) and luminosities. As they study the H-R diagram, astronomers use it to piece together the complex story of how stars are born, live their lives and die.

On the H-R diagram, most stars lie on a diagonal band called the main sequence that links hot, bright blue stars at the upper left with cool, dim red stars at the lower right. Scientists believe that all these stars are burning energy in the same way as the Sun – the nuclear fusion of hydrogen atoms into helium. Differences in colour and brightness depend on a star's supply of hydrogen in its core and, therefore, also on its mass. Bright blue stars are the most massive, and dim red stars are comparatively lightweight. Because stars rarely gain or lose mass, and their hydrogen is used up at a steady rate, most of them stay at almost the same point on the main sequence for most of their lives. Stars that are not on the main sequence have either not begun to burn hydrogen or are so old that they have used it all up.

The physical differences in stars at different points on the diagram can be seen in their colour, which indicates the surface temperature. Redder stars are cooler, so less energy must be escaping from their surface. Bluer stars are hotter, so more energy is escaping. If two stars have the same luminosity but different colours, then they must be different sizes.

Stellar life cycles This H-R diagram shows the paths taken by stars of different masses after they leave the main sequence. A Sunlike star swells to become a red giant, shrinks and swells again, before puffing off its outer layers as a planetary nebula and shrinking to become a white dwarf. A blue star of eight solar masses swells to a red supergiant and goes through several pulses of shrinking and expansion before blowing itself apart in a supernova explosion.

They are generating the same amount of energy, but the larger one has a greater surface area for radiation to escape from, so it is cooler. A highly luminous star that appears red must be huge, but a dim star that still appears white must be incredibly compressed. Most stars, however, sit on the main sequence and get bluer as they get brighter.

Because the H-R diagram tells us about stellar sizes, it can also help astronomers to work out what is going on inside stars at different points in the diagram. All stars are poised in a delicate equilibrium between the outward pressure of radiation and the inward pull of their own gravity. A relatively small change in the amount of energy a star generates can cause the star to swell, or to shrink to a fraction of its previous size. So a star that grows brighter or dimmer due to a change in its energy source will also change colour.

Astronomers think that during a star's lifetime, a long period of hydrogen-powered life on the main sequence is preceded and followed by briefer periods of rapid change. Because physicists know the amount of energy emitted by a single nuclear fusion reaction, they can use the overall energy emitted by a star to calculate the rate at which its core is using hydrogen and then work out its life expectancy.

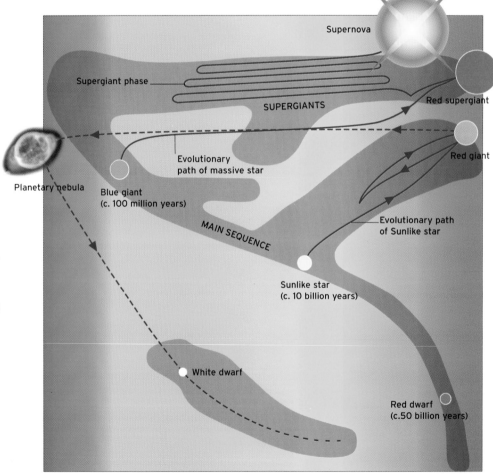

Supernova

Supergiant phase

SUPERGIANTS

Red supergiant

Evolutionary path of massive star

Red giant

Planetary nebula

Blue giant (c. 100 million years)

MAIN SEQUENCE

Evolutionary path of Sunlike star

Sunlike star (c. 10 billion years)

White dwarf

Red dwarf (c.50 billion years)

See also: Star colours 88 · Plotting the lives of stars 90 · Red giants 112 · White dwarfs 118 · Supernovae 120

The lives of Castor and Pollux

α Gem Alpha Geminorum, or Castor, has six components. To the naked eye, it shines at magnitude 1.6. However, a small telescope reveals two components of magnitudes 1.6 and 2.6, each white and farther up the main sequence than the Sun, with a far fainter red-dwarf companion. Each of these stars is itself a double, too close to separate visually. The entire system lies around 52 light-years away.

18 GEMINI *The constellation Gemini, the Twins, is one of the brightest winter constellations, best seen around December and January. The constellation is named after the twin stars Castor and Pollux, similar in size and arranged together at the head of the constellation. Another pair of slightly fainter stars lies parallel to Castor and Pollux, forming a roughly rectangular shape. Gemini is a zodiac constellation, crossed by the star clouds of the outer Milky Way, and is rich in star clusters and interesting stars. The bright stars of Gemini also show a wide range of different stages in stellar evolution. One of the year's best meteor showers, the Geminids, also radiates from here.*

η Gem Eta Geminorum is a red giant with a faint companion. It is also variable. It has swollen so much that it has become unstable, and pulsates between magnitude 3.2 and 3.9 in just under eight months. Eta is around 350 light-years away, much farther away than Castor or Pollux.

β Gem Beta Geminorum, or Pollux, is actually the brightest star in Gemini, shining at magnitude 1.2, but was wrongly labelled Beta in 1603. Pollux is a much simpler star than Castor. It is a yellow giant around 34 light-years away, evolving toward its red-giant phase.

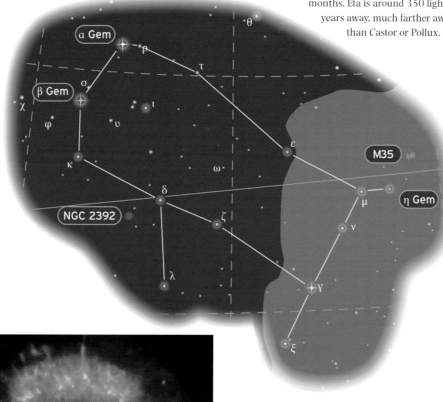

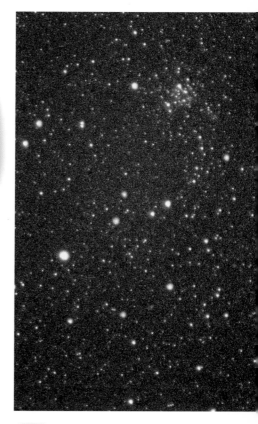

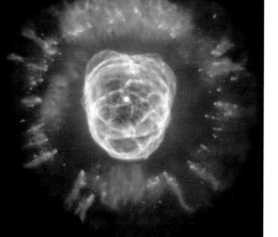

NGC 2392 The Eskimo planetary nebula is so called because it seems to show a face in a furry hood when viewed through high-power telescopes. It appears as a fuzzy blue-green star of magnitude 9.0 through a smaller instrument at low magnification. Higher magnification may reveal the faint central star surrounded by a fuzzy ring half a light-year across. The Eskimo Nebula lies around 2,300 light-years from Earth.

M35 The open star cluster M35 contains about 200 stars, and lies around 2,800 light-years away. It covers half a degree in the sky (the same size as the Full Moon) and is visible in dark skies with the naked eye, shining at magnitude 5.6. The cluster's brightest stars are blue-white and yellow-orange giants, roughly 400 times as luminous as the Sun.

94 Star birth

Stars are born from great protostellar nebulae – clouds of gas and dust hundreds of light-years across. Over millions of years, these clouds slowly collapse and separate into individual centres known as protostars, from which stars are formed.

Interstellar space is rich in gas and dust clouds – mixtures of primordial hydrogen and helium and more complex elements formed by earlier generations of stars. As these ancient stars came to the end of their life cycles, they scattered their remains across nearby space.

Normally, these cold, dark clouds hang undisturbed in space and remain invisible except where they are silhouetted against a brighter background – as in Orion's famous Horsehead Nebula. Occasionally, however, they are disturbed, and, because they are relatively dense, they begin to contract under their own gravity. Disturbances that trigger waves of star formation can be caused by shock waves from a nearby supernova explosion, by the passage of a runaway star, or by waves of compression that regularly circle within a spiral galaxy.

Dark protostellar clouds come in two types – large and irregular Barnard Objects, tens of light years across, and smaller, spherical Bok Globules. As either type of cloud begins to collapse, it will usually fragment into smaller clumps. These clumps, which are several times the size of the Solar System, become the seeds of individual stars. The mass of material within these protostars can vary wildly, giving rise to anything from a tiny dwarf star to a massive giant. As the protostar slowly collapses inward, its temperature and internal pressure increase. As its interior grows hotter and denser, it radiates huge amounts of energy, but is so vast that it remains relatively cool, and emits only red or infrared light. In fact, protostars are rarely seen in normal light, since they are still hidden within their dark nebulae. Only infrared (and radio) radiation can lift this veil and reveal the newborn stars that lie within.

The speed of a protostar's collapse depends on its gravity and mass, and the collapse only stops when the temperature at the star's core is hot enough to trigger nuclear reactions. The few protostars with masses greater than 100 Suns reach such high temperatures and pressures so rapidly that they blast themselves apart and never reach the main sequence. Typical giant stars take around 20,000 years to start shining, while stars such as the Sun take millions of years to reach the same stage. In both cases, as nuclear reactions kick in, the outward pressure they generate through the star halts the gravitational collapse.

Stellar nursery The glowing heart of the Orion Nebula, M42, is home to four young stars, known as the Trapezium. Ultraviolet radiation from these stars excites hydrogen atoms in the nebula, causing it to glow.

Emerging stars A cluster of brilliant young stars are emerging from the top of a dark cloud of gas and dust in which they formed.

Chain reaction Hot gases are blasted off from the young stars, triggering the formation of more stars inside the nebula cloud.

Cloud meltdown As the process of star formation continues, the nebula is either compressed into stars or blasted into space.

See also: Genesis 34 · Star colours 88 · Stellar evolution 92 · Red & brown dwarfs 102 · Stellar heavyweights 104

Star formation in Orion

26 **ORION** *Representing the Hunter, Orion is one of the most recognizable and spectacular constellations. Bright stars mark his shoulders and knees, and a chain of three stars, Orion's belt, straddle his waist. The bright nebula M42 marks where a sword hangs from the hunter's belt, and even Orion's raised club and shield are easily identifiable. Orion also contains a huge stellar factory, where young stars are born behind dark dust and gas clouds. Only a few of these stars are able to shine unhindered by the dust, but their strong light causes the clouds around them to glow, creating emission nebulae that are visible even with the unaided eye.*

α Ori Betelgeuse, which marks Orion's shoulder, is one of the sky's most prominent red giants – an enormous star nearing the end of its life which has swollen to more than 800 times the diameter of the Sun. Betelgeuse lies 427 light-years away, and is so large that astronomers have managed to map its surface, producing images of huge dark spots on its surface.

ζ Ori Zeta Orionis, a blue-white star of magnitude 1.7, lies 815 light-years away. Its light illuminates an emission nebula, in front of which lies the dark cloud Barnard 33, better known as the Horsehead Nebula. Unfortunately, although it photographs spectacularly in large telescopes, the Horsehead is extremely faint and beyond the reach of most amateur instruments.

M42 The centrepiece of Orion is the Great Nebula, M42 but as shown in this widefield picture of the whole Orion complex, several other nebulae lie close to it in the sky. At the heart of M42, 1,500 light-years away, lies a young quadruple star known as the Trapezium. Its members have brightnesses ranging from magnitude 5.4 to magnitude 7.0, and two of them are variable eclipsing binaries. The Trapezium is a very rewarding target for a small telescope.

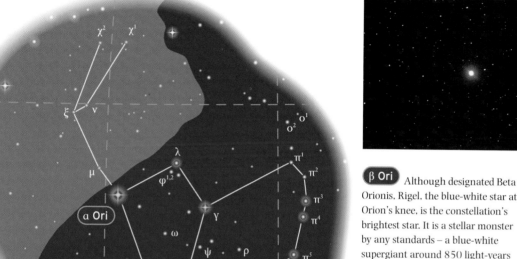

β Ori Although designated Beta Orionis, Rigel, the blue-white star at Orion's knee, is the constellation's brightest star. It is a stellar monster by any standards – a blue-white supergiant around 850 light-years from Earth and nearly 50,000 times as bright as the Sun.

δ Ori Delta Orionis in Orion's belt is a blue-white magnitude 2.2 star. Situated around 920 light-years away, it is part of the same star-forming complex, M42, that produced most of Orion's stars. Binoculars suggest that it has a companion of magnitude 6.9, but this is just a line-of-sight effect. However, Delta does have a genuine companion star, which orbits and eclipses it every 5.7 days, causing the star to dim slightly.

STAR STORIES

Orion and the Pyramids

Orion was very significant to the ancient Egyptians, and there have been many theories linking the pyramids of Giza to this constellation. One idea was that the layout of the three pyramids reproduces the pattern of Orion's belt. There is more solid evidence that diagonal shafts leading from the burial chamber are aligned with Orion.

96 Young stars

A star's coming of age – the point when it begins to shine by nuclear reactions rather than gravitational contraction – is a violent time. Most young stars blow away the remnants of the dark nebula that formed them, sometimes causing new stars to form.

By the time it joins the main sequence of hydrogen-burning stars, a protostar may still be surrounded by gas with two or three times its own mass. Young stars lose this mass in two main ways, and most of the excess is blown away in a million years or less. Stars with the mass of three Suns or less go through a stage of rapid fluctuation called the T Tauri phase (named after the first such star discovered) and all stars go through a process called bipolar outflow, in which mass is flung out from the star's poles in huge jets.

Just before a star ignites, it passes through its smallest and densest phase – with no nuclear reactions to hold it up, it shrinks under its own weight to just a few times the size of Jupiter. The incredible pressure and temperature this generates is the trigger that starts nuclear reactions. As radiation pushes out through the star, it expands again until it reaches a point where the outward pressure of radiation balances the inward force of gravity.

The T Tauri phase happens as the star is trying to find this equilibrium point. At first it expands too far, blowing its outer layers off into space, then it contracts again, bouncing back and forth about the balance point until it finally stabilizes. As shells of gas are blown out from the star at speeds of around 300,000 km/h (185,000 mph), they block out some of the star's light, causing it to vary in brightness in just a few days. The T Tauri phase of a star's evolution lasts around 10 million years and can result in the loss of a solar mass of material. More massive stars also go through a similar process as their fierce radiation blows excess gas away, but they do not vary in brightness.

The second type of mass loss, bipolar outflow, lasts just a few thousand years but ejects more mass than ends up in the final star itself. As the protostar collapses inward, the random motion of the gas cloud is converted into rotation, which flattens the cloud into a broad disc surrounding the still-forming star. As particles within the disc jostle each other, they lose energy and spiral in toward the growing star in a process called accretion. However, most of the material from the accretion disc is not absorbed into the star. Instead it is swept up and ejected at hundreds of kilometres per second in bipolar jets along the star's axis of rotation. These jets moderate the rotation of the young star and prevent it spinning itself to pieces. As the bipolar jets collide with nearby gases, they heat up and create glowing knots called Herbig-Haro (HH) objects.

Youthful exuberance The pinky glow of a Herbig-Haro object (right), heated up by the bipolar jets of Gamma Cassiopeiae, betrays the star's youth. An artist's impression of the young star (above) shows the accretion disc around it and the jets of gas streaming out from either pole.

See also: Genesis 34 · Star distances 84 · Star colours 88 · Stellar evolution 92 · Star birth 94

The Ethiopian Queen

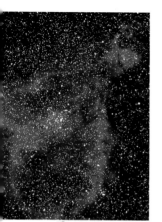

IC 1805 Containing about 40 stars in an area slightly smaller than the Full Moon, IC 1805 is a bright star cluster, 6,850 light-years away, which shines at a magnitude of 6.5. It makes an attractive sight in binoculars or a telescope at low magnification, and is surrounded by a faint emission nebula only visible through large telescopes.

γ Cas Gamma Cassiopeiae, the young and violent blue-white star at the heart of the Cassiopeian W, reveals its secret through variations in brightness. Normally, the star has a magnitude around 2.5, but as it ejects gas it can vary from magnitude 3.0 to 1.6. It lies 600 light-years away.

10 CASSIOPEIA *The five brightest stars of Cassiopeia form a distinctive W-shape that circles the pole star on the opposite side to the Plough, making it among the most recognizable of the northern constellations. The five stars of the W lie against the Milky Way, and the young star Gamma forms the point at their centre. Cassiopeia is rich in star clusters and nebulae, as well as being the site of the strongest radio source in the sky, Cassiopeia A – an expanding shell of gas left by a supernova 10,000 light-years away.*

M52 M52 is a small but rich cluster of approximately 100 stars, which is about one-third of a Moon-width across and 4,800 light-years from Earth. Most of the stars in the cluster have magnitudes of 8.2 or less, but their combined light gives the cluster a total brightness of magnitude 6.9, so it can be seen in binoculars. A magnitude 5.0 yellow star in the foreground, 4 Cassiopeiae, helps mark the cluster's position.

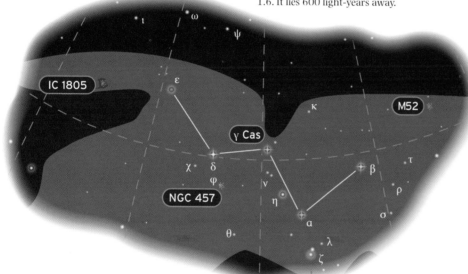

NGC 457 Situated 9,100 light-years away, NGC 457 is Cassiopeia's brightest star cluster. It contains 80 stars and is just visible to the naked eye. The stars of the cluster are arranged in chains and loops that reveal the structure of the protostellar cloud from which they formed. Phi Cassiopeiae, a magnitude 5.0 yellow supergiant, lies in the middle of the cluster, but is most likely a foreground star.

STAR STORIES

Cassiopeia

In Greek legend, Cassiopeia was a queen of Ethiopia who incurred the anger of Hera, queen of the gods, by boasting of her daughter Andromeda's beauty. A sea monster, Cetus, was sent to ravage the kingdom, and could only be appeased by the sacrifice of the princess. She was only saved from the monster by Perseus on his horse Pegasus. The distinctive W-shape depicts Cassiopeia on her throne.

OTHER YOUNG STARS

NAME	CONSTELLATION	MAGNITUDE
48 Librae	Libra	4.9
Pleione	Taurus (Pleiades cluster)	4.8–5.5
T Tauri	Taurus	9.8/13.5
R Coronae Australis	Corona Australis	5.7/14.8
Delta Scorpii	Scorpius	2.3
S Doradus	Dorado	8.6/11.5

Sunlike stars

The evolution of a star such as our Sun is a long and stately process, typical of how most stars in the Universe evolve. It will remain on the main sequence for billions of years while it consumes the hydrogen fuel in its core.

All stars on the main sequence shine from the energy generated as they convert hydrogen into helium at their cores, but there are several different processes – dependent on the mass of the star – by which this may happen. In stars of similar mass to the Sun, energy generation is dominated by a fairly simple process called the proton-proton chain. To understand this requires some knowledge of the internal structures of hydrogen and helium atoms.

Hydrogen is the simplest element of all. It has a central nucleus of just one particle, a positively charged proton, around which orbits a much lighter, negatively charged electron. The high temperatures within a star strip away the electrons from the hydrogen nuclei, leaving the protons exposed. Helium is the next-simplest element, but it is more

complex. Its nucleus is made up of four particles – two protons, and two uncharged neutrons of roughly equal mass. The two electrons which orbit a helium atom are also stripped away inside a star, laying the nucleus bare. Nuclear fusion combines four hydrogen nuclei to make one helium nucleus. In the process, two protons are transmuted into neutrons, several new particles are produced, and a large amount of energy is released.

Where does the energy released by fusion actually come from? In each of the three proton-proton chain reactions, the combined products have less mass than the particles which reacted to form them. The change in mass is tiny (less than one-thirtieth the mass of a single proton) but it is converted directly into energy according to Einstein's equation $E = mc^2$

(where c is the speed of light). In a star such as the Sun, several million tons of material are converted into energy each second, generating around 390 trillion megawatts of power. Only about 10 per cent of the star is hot enough for fusion, but that can total several hundred trillion trillion tons – enough to allow fusion to continue for billions of years.

During a star's main sequence lifetime, the density of its core increases as a large number of hydrogen atoms are replaced by far fewer helium atoms. Convection keeps the core well mixed, pumping enough hydrogen into the centre for fusion to continue, but the increased pressure gradually causes the core's temperatures to rise, until it begins to affect the hydrogen surrounding it. Eventually the temperature in a layer around the core becomes high enough for hydrogen fusion to begin burning in a thin shell surrounding the core. When the fuel supply in the core finally falters and dies, this burning shell will become the star's main source of energy.

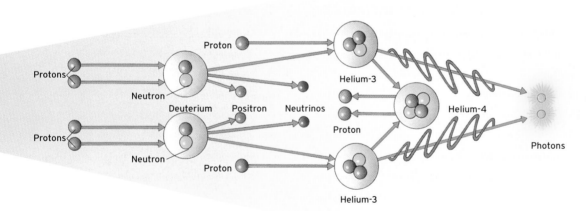

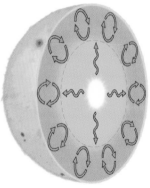

Nuclear fusion In the proton-proton nuclear fusion chain, four hydrogen nuclei combine to form one helium nucleus. But the chances of four protons colliding simultaneously are extremely slim, so fusion happens by a three-step process. First, two protons combine to form deuterium, or heavy hydrogen. An extra proton makes helium-3, which, when combined with another helium-3 atom, makes helium-4 and releases two excess protons.

Inside a Sunlike star A star with similar mass to our Sun transports energy outward by radiation close to the core. Once the temperature has dropped low enough for bulk movements of gas, heat is dissipated by convection.

See also: Plotting the lives of stars 90 · Stellar evolution 92 · Red and brown dwarfs 102 · Stellar heavyweights 104

The herdsman's suns

μ Boo Mu Boötis marks the herdsman's staff. It is a triple star 120 light-years away. To the naked eye it is a blue-white, magnitude 4.3 star, but it has a wide companion of magnitude 6.5, which is split into two stars of magnitudes 7.0 and 7.6.

ε Boo Epsilon Boötis is a double star known as Pulcherrima – the Most Beautiful. A small telescope will split the orange star to give a bright, orange giant of magnitude 2.4 with a blue companion of magnitude 4.2, both 200 light-years away.

5 BOÖTES *The bright constellation Boötes is a northern circumpolar constellation lying to the east of Ursa Major. It represents a celestial herdsman driving the Great and Little bears around the sky. The constellation is easily spotted, because it contains Arcturus – the Bearkeeper – the fourth-brightest star in the sky. Elsewhere, Boötes is a large but relatively average constellation – despite lying west of the bright Milky Way star fields, it is not rich in galaxies. However, Boötes does contain several interesting stars, and is the source of the year's richest and most reliable meteor shower, the Quadrantids – named after a long-lost constellation. These peak in early January each year with 100 meteors an hour.*

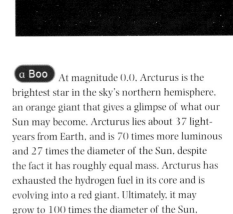

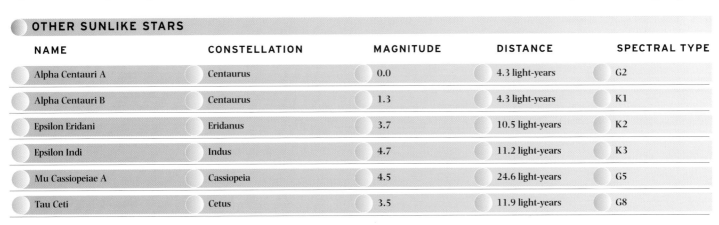

τ Boo Tau Boötis is a nearby Sunlike star still on the main sequence. It lies just 51 light-years from Earth, and is slightly hotter and 25 per cent more massive than the Sun. It also has at least one planet in orbit around it. However, this world has a mass at least three and a half times greater than Jupiter, and circles its sun in just 3.3 days. Tau Boötis's planet was the first to have its light directly detected and separated from its parent star, revealing its blue-green colour.

ξ Boo Xi Boötis is another attractive double star for small telescopes. It lies just 22 light-years away, and consists of a yellow, magnitude 4.7 star and an orange, magnitude 6.8 star, locked in a 150-year orbit around each other.

α Boo At magnitude 0.0, Arcturus is the brightest star in the sky's northern hemisphere, an orange giant that gives a glimpse of what our Sun may become. Arcturus lies about 37 light-years from Earth, and is 70 times more luminous and 27 times the diameter of the Sun, despite the fact it has roughly equal mass. Arcturus has exhausted the hydrogen fuel in its core and is evolving into a red giant. Ultimately, it may grow to 100 times the diameter of the Sun, and become the brightest star in the night sky.

OTHER SUNLIKE STARS

NAME	CONSTELLATION	MAGNITUDE	DISTANCE	SPECTRAL TYPE
Alpha Centauri A	Centaurus	0.0	4.3 light-years	G2
Alpha Centauri B	Centaurus	1.3	4.3 light-years	K1
Epsilon Eridani	Eridanus	3.7	10.5 light-years	K2
Epsilon Indi	Indus	4.7	11.2 light-years	K3
Mu Cassiopeiae A	Cassiopeia	4.5	24.6 light-years	G5
Tau Ceti	Cetus	3.5	11.9 light-years	G8

100 Other solar systems

Until recently, astronomers had searched in vain for evidence of planets orbiting other stars. Some thought our Solar System was unique in the Universe. Since the early 1990s, however, new techniques have revealed dozens of strange new worlds.

A solar system in the making? The fledgling planetary system belonging to the star Beta Pictoris in the constellation Pictor seen as a flat disc of gas and dust thousands of billions of kilometres in diameter.

A star is born from a collapsing cloud of interstellar gas and dust. As it collapses, the cloud forms a flattened disc of dust and gas around the newborn star. Such a protoplanetary disc gave birth to our Solar System. Astronomers have detected many of these discs surrounding stars at an early stage of their formation. The next stage – the development of discs of warm material around young but fully fledged stars such as Beta Pictoris – has also been observed. But, until recently, astronomers have had no luck in detecting actual planets.

The main difficulty in detecting worlds orbiting other stars is their faintness. Even the largest and brightest planet will be outshone around one billion times by its own sun. As a consequence, any planets are simply lost in the glare.

Fortunately, it is possible to find planets indirectly – by measuring the effect on their parent star. In theory, this can be done in three ways. One is to plot the movements of a star in the sky with extreme precision, revealing any wobble caused by the gravitational tug of a planet. Another is to watch for tiny dips in the brightness of a star, caused by a planet passing in front and eclipsing it.

The third method has proved by far the most successful. Starlight from target stars is split into a spectrum and scanned for tiny shifts in the positions of its dark absorption lines. Such Doppler shifts are caused as the star's wobble moves it toward and away from Earth.

Several dozen extrasolar planets, with masses equal to or greater than Jupiter's, have now been detected, including a handful of multiple-planet systems. These new worlds have several features that surprised astronomers familiar with our own Solar System. Many of them orbit very close to their stars (in several cases, even closer than Mercury's orbit around our Sun) in a zone where no gas-giant planet should ever form. Most of them also have markedly stretched or elliptical orbits, very different from the near-circular orbits of planets in our home system.

These discoveries are forcing astronomers to reconsider their models of how solar systems form. It now seems that near-circular orbits are far from the norm, and giant planets can slowly spiral in toward their stars, disrupting any smaller planets whose orbits they cross, and finally being swallowed up by the star itself.

New planets are being discovered all the time, but current detection methods are biased toward finding massive planets in short-period orbits. So far the only planets of Earth-like masses have been discovered around pulsars— rapidly spinning dead stars, whose precise spin periods can be affected by the pull of planets orbiting them. A new generation of telescopes in space may soon change this, however. Astronomers will be able to monitor stars over longer periods and more precisely, so they will start finding smaller planets, plus those in longer-period orbits.

See also: The home system 32 · Genesis 34 · Stellar evolution 92 · Sunlike stars 98 · Astronomy from space 200

Planet hunting in Pictor

73 PICTOR *This dim constellation lies at mid-southern latitudes, midway between the brilliant star Canopus (in Carina), and the Large Magellanic Cloud. Pictor was described by Lacaille, who originally called it Equuleus Pictoris – the Painter's Easel. Beta Pictoris, the constellation's brightest star, and of interest to planet hunters, lies at its northern tip.*

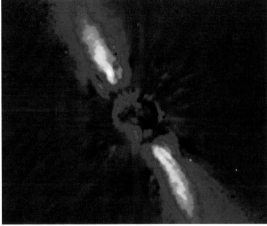

γ Pic Gamma Pictoris, a yellow star of magnitude 4.5, is a bright giant star 173 light-years away.

δ Pic Delta Pictoris is a blue, eclipsing binary star, 1,650 light-years away with brightness varying between magnitude 4.7 and 4.9 every 40 hours. The stars are so close together that their gravity distorts them into egg shapes.

Kapteyn's Star This is the second-fastest-moving star in the sky. It is a magnitude 8.9 red dwarf, 12.8 light-years away and visible only through binoculars. It bears the name of the Dutch astronomer Jacobus Kapteyn, who discovered its rapid movement in 1897.

ι Pic Iota Pictoris is a double star easily distinguished in a small telescope. It consists of two stars around magnitude 6.0, orbiting each other 120 light-years away.

β Pic This image shows a planet-forming dust disc around the magnitude 3.9 star Beta Pictoris. The image, made at a single wavelength and with the star itself blocked out, shows that the dim circumstellar disc is warped close to the star, hinting at the presence of at least one planet. Beta Pictoris lies 63 light-years from Earth.

✦ STAR STORIES

Jacobus Kapteyn

Kapteyn (1851-1922) was a Dutch astronomer who did the first extensive photographic survey of the southern skies. He hoped to find nearby stars by identifying those with large parallaxes. By accident, he found one dim star was moving across the sky fast enough to span the Full Moon's diameter every 200 years. His findings with star movements led him to discover star streaming - the fact that stars in one part of the sky appear to move in one direction; and those in another part, the opposite way. This was the first evidence that the Milky Way is rotating.

SOLAR SYSTEMS AROUND SUNLIKE STARS

NAME	CONSTELLATION	PLANET MASS	DISTANCE FROM STAR	ORBITAL PERIOD	DATE FOUND
51 Pegasi b	Pegasus	0.5 Jupiters	0.05 AU	4.2 days	1995
70 Virginis b	Virgo	6.6 Jupiters	0.26–0.60 AU	116.6 days	1996
Rho Coronae Borealis b	Corona Borealis	1.1 Jupiters	0.23 AU	39.6 days	1997
Tau Boötis b	Boötes	3.9 Jupiters	0.05 AU	3.3 days	1997
Epsilon Eridani b	Eridanus	0.9 Jupiters	1.30–5.30 AU	6.9 years	2000
HD 74156 b, c	Hydra	1.6, >7.5 Jupiters	0.10–0.46, 2.70–6.24 AU	51.6 days, 6.3 years	2001

102 Red and brown dwarfs

We see only the brightest stars in the galaxy—usually those with the mass of the Sun or more. But most stars are invisible to even the most powerful telescopes. These are the dim red dwarfs and the brown dwarfs, failed stars too small to shine properly.

Until recently, astronomers could only speculate about just how many dwarf stars our galaxy might contain. Even the closest dwarf – Proxima Centauri – is invisible to the naked eye at magnitude 11.0, and, beyond a certain distance, dwarfs were simply impossible to detect. However, recent advances in infrared (heat) astronomy have revealed that our galactic neighbourhood contains many other red and brown dwarfs, and there is no reason to think that the rest of the galaxy is any different.

Red dwarfs typically have less than 80 per cent of the mass of the Sun. Their low mass means they have much lower temperatures and pressures in their cores but, provided they weigh more than 8 per cent of the mass of the Sun, they can still generate energy through normal nuclear reactions.

Although dwarfs are dimmer than the Sun, they are not much smaller, because their gravity has less effect and outward pressure from radiation and convection helps to counterbalance the tendency to collapse. For example, a 0.2 solar-mass star may still have a diameter one-third of the Sun's, even though it may shine with only 1 per cent of the Sun's light.

Below the cutoff point for normal nuclear fusion – 8 per cent of solar mass – lies another type of faint object, intermediate between a star and a giant planet, called a brown dwarf.

Brown dwarfs are giant balls of gas between 12 and 80 times the mass of Jupiter – too small to shine by the proton-proton process of hydrogen fusion, but massive enough to trigger another nuclear reaction that requires far lower temperatures – the fusion of deuterium. Brown dwarfs also generate substantial amounts of energy as they contract under their own gravity.

Brown dwarfs are easiest to detect when they orbit another star, since their gravity slightly perturbs the movement of the brighter companion. But they are not simply oversized giant planets. They form from collapsing gas clouds in the same way as normal stars, and dozens of free-floating brown dwarfs have now been detected from their infrared (heat) emissions in star-forming regions such as the Orion Nebula. In fact, protostellar nebulae probably create just as many brown dwarfs as more massive stars.

In theory, the largest giant planets can weigh more than the smallest brown dwarfs – the two objects are distinguished by the way they form rather than by their mass. But although the heaviest brown dwarfs weigh far more than gas-giant planets, they should not be significantly larger, because their increased gravity causes them to compress further. Gliese 229B, in Lepus, weighs 40 times as much as Jupiter, despite having a very similar diameter. Brown dwarfs on the verge of becoming shining stars are even smaller.

Dwarfs discovered Astronomers had predicted the existence of brown dwarfs for some time, but their faintness made them extremely difficult to discover. The first one to be found was Gliese 229B, shown in this artist's impression, which was orbiting a faint star in the constellation Lepus, the Hare, 18 light-years from Earth.

Inside a dwarf star The low energy output of red dwarfs means that they tend to have a simpler structure than Sunlike stars - heat is passed from the core to the surface by convection in cells of rising hot gas, without the radiative zone of larger stars.

See also: Stellar evolution 92 · Sunlike stars 98 · Stellar heavyweights 104 · White dwarfs 118 · The invisible Universe 182

Dwarfs in Lepus

62 LEPUS *The Hare is a small constellation lying directly to the south of Orion and east of Canis Major. The name dates back to ancient Greek times, and features in the complex Orion legend. Lying some way to the south of the Milky Way, Lepus contains the globular cluster M79, associated with our own galaxy, and other more distant galaxies that require a large telescope to see. The central stars of the constellation have a distinctive bow-tie shape. The red-dwarf star Gliese 229B, with its brown dwarf companion, lies in the constellation's southwest corner, but even its parent star is invisible except with a powerful telescope.*

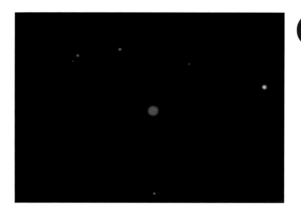

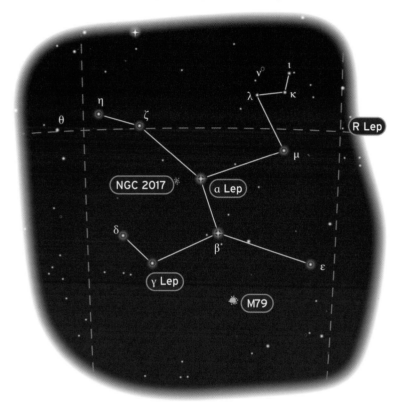

R Lep The deep-red star R Leporis, also called Hind's Crimson Star, is a long-period variable similar to Mira in Cetus. It is an unstable star, pulsating in size with a 430-day period and varying in brightness from magnitude 5.5 to magnitude 12.0.

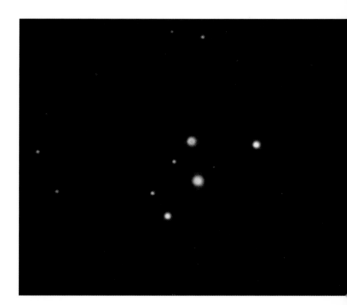

NGC 2017 This is a beautiful multiple star, with five members of magnitudes 6 to 10 visible in a small telescope. Two of these stars are double, and there is a very faint eighth star in the group as well.

γ Lep Gamma Leporis is a nearby and easily seen double star, just 29 light-years away. It consists of a yellow, magnitude 3.6 star orbiting with an orange, magnitude 6.2 companion.

α Lep Alpha Leporis is a white supergiant star, 1,280 light-years away and shining at magnitude 2.6.

M79 This is one of the best globular clusters visible from the northern hemisphere. At magnitude 7.7, this huge ball of stars 42,000 light-years away is easily seen with binoculars as a fuzzy star one-quarter the size of the Full Moon. A large telescope will separate the individual stars.

Stellar heavyweights

High-mass stars live life in the fast lane. Their cores reach such extreme temperatures and pressures that they burn brighter and hotter than all other main-sequence stars. They also die spectacularly, destroyed in enormous supernovae explosions.

Stars with masses several times that of the Sun lead brief but brilliant lives. While on the main sequence, the high mass of these stars opens the way for more powerful nuclear reactions in their cores that use heavy elements to catalyse the fusion of hydrogen into helium. This process releases far more energy than normal fusion, but only works efficiently at temperatures of 20 million °C (36 million °F), so they spend most of their lifetimes as bright blue or white O-type stars—the bluest and hottest star type. Often, such high-mass stars are called blue giants – they are bigger than any other main sequence stars. However, they are still dwarfed by red giants and supergiants which represent later stages in a star's life cycle.

The most massive of high-mass stars also have the energy to convert helium into heavier elements, such as carbon, while on the main sequence. This is unusual since stars tend to produce heavy elements much later in their lives, and it makes these stars factories for the production of heavy elements. The enormous explosions that end their lives scatter such elements across space.

The penalty for the huge amount of energy generated in giant stars is a shortened lifetime. Hydrogen is squandered rapidly, so a typical giant star of eight solar masses survives for only 100 million years on the main sequence before the hydrogen in its core is used up. In contrast, a Sunlike star can last for 10 billion years.

The main sequence stage is even shorter for the most massive stars because of another effect. The fierce radiation generated by the star creates a powerful stellar wind that rapidly blows away the star's outer layers. In this way, a monster star that starts its life at 100 solar masses will lose more than half its material in its brief million years on the main sequence. Finally, the star's outer layers are shed, and the star's core, now rich in heavy elements, is exposed. The result is an extremely hot Wolf-Rayet star, with surface temperatures that can reach 50,000 °C (90,000 °F).

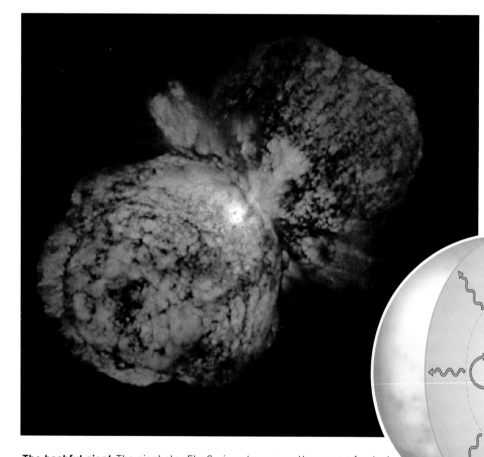

The bashful giant The giant star Eta Carinae hovers on the verge of naked-eye visibility at around magnitude 6.0. It is incredibly brilliant, but lies hidden amid clouds of dust and gas thrown off from its outer layers. Eta, however, is an unpredictable variable star that grew to rival Canopus in 1843 and could brighten again.

Inside a stellar heavyweight
Massive stars have extremely hot cores, but their surface temperatures are just a few thousand degrees hotter than the Sun. The large temperature difference gives them a very different internal structure. The central regions around the core move energy upward by convection, and it is then radiated through the outer layers - exactly the opposite of what happens in a Sunlike star.

See also: Stellar evolution 92 · Sunlike stars 98 · Red giants 112 · Pulsating stars 114 · Supernovae 120

The sky's megastars

44 **CARINA** *Known as the Ship's Keel, this is one of the largest and brightest southern constellations, dominated by Canopus, the second brightest star in the sky. The constellation lies close to the southern celestial pole, stretching nearly a quarter of the way around it, and never sets for most of the southern hemisphere. Carina is rich in interesting stars, clusters and nebulae because it lies in one of the densest parts of the Milky Way. The constellation rises high in the sky on southern summer nights, and Canopus appears over the winter horizon of the southern United States. North of Theta Carinae and its surrounding cluster of stars lies the Carina Nebula, one of the night sky's most spectacular objects.*

NGC 3372 Also known as the Carina Nebula, this is a spectacular group of four bright gas clouds separated by a dark nebula – the Keyhole Nebula. The separate clouds are easily visible to the naked eye, and the whole nebula is more than four Moon-widths across. It lies 9,000 light-years away, surrounding the giant star Eta Carinae, and shines at magnitude 6.5.

ι Car The white supergiant Iota Carinae, also known as Turais, lies 690 light-years away, and shines at magnitude 2.2.

α Car The brilliant white star, Alpha Carinae or Canopus shines at magnitude -0.6. It is only outshone by Sirius – but while the Dog Star is relatively nearby at just over 8 light-years, Canopus is a distant supergiant, more than 300 light-years away.

NGC 2516 Visible with the naked eye at overall magnitude 3.3, this cluster of 80 stars lies 1,400 light-years away and is a Full Moon's width. At the centre is a red giant, magnitude 5.2, which can be seen by using a small telescope.

IC 2602 This is a large cluster of young stars, 500 light-years away, and centred on Theta Carinae – magnitude 2.7. The cluster contains 60 stars, many of which are visible to the naked eye, and is twice the width of the Full Moon.

υ Car The white star Upsilon Carinae is, in fact, a double star 1,600 light-years away. A small telescope separates it into separate components – a magnitude 3.0 white supergiant and a magnitude 6.0 blue star.

STAR STORIES

The Lost Constellation

Carina (the Ship's Keel) and its neighbouring constellations Puppis (the Stern) and Vela (the Sail), were once members of a single, much larger constellation - Argo Navis or the Ship - named after the vessel the Greek hero Jason used on his quest for the Golden Fleece. Dating back to ancient times, the constellation was broken apart in the 1750s by French astronomer Nicolas Louis de Lacaille who thought it was just too huge and cumbersome to deal with. The pieces remain, though, along with Columba, the Dove that guided Jason through the Dardanelles and into the Black Sea.

Multiple stars

A casual glance at the night sky shows that it seems to be full of single stars, similar to our Sun. But in fact most of our galaxy's stars belong to double- and multiple-star systems, containing two, three or even more stars orbiting each other.

Multiple stars begin their lives in the same way as single stars, within collapsing clouds of gas and dust. Instabilities cause the cloud to separate into smaller chunks in orbit around each other and the stars continue to collapse until they join the main sequence and begin to shine. Although systems of three or more stars are not uncommon, double or binary systems seem to form most easily. They account for more than half the stars in our galaxy, and many larger systems are built up from close, double-star pairs.

Binary stars orbit each other at a variety of distances. Those with the largest separations may take millions of years to circle one another, while at the other extreme, some stars spin around each other in a few days or even hours. Both stars orbit the system's centre of mass – a point somewhere in between them. If the two stars in the system are of equal mass, then the centre is half-way between them, but if one is more massive than the other, this point will be closer to the heavier star. Knowing the position of the centre of mass allows astronomers to work out the stars' relative weights. This is the only way that astronomers can weigh stars and it helps to confirm that a star's brightness on the main sequence depends directly on its mass, and also that stars of different masses evolve at different rates.

A binary system may be made up of a Sunlike star with billions of years of life ahead of it, and a supermassive star already in its death throes, or even the dead remnants of a star. Yet the pair must have formed at the same time.

Although telescopes can show the components of many binary stars, most are either so far away or so tightly bound that they are impossible to see individually. They can be revealed in other ways, however. For example, some binaries can be detected when their orbits line up so that they pass directly in front of each other when seen from Earth, creating an eclipsing binary (a star that periodically dips in brightness). Another way of detecting a close binary is to separate its light into a spectrum. When astronomers see a star's spectral lines doubled up, they know they are detecting two stars that appear as one, but whose spectral lines are periodically shifted apart by the Doppler effect when one moves toward, and the other away from us. Such star systems are known as spectroscopic binaries.

In some binaries, one rapidly ageing star may swell into a giant whose outer layers come within the gravitational grasp of its companion. In such contact binaries, material moving between stars can affect their evolution and create strange eclipsing variables.

Gaseous ring The Beta Lyrae system, shown in this artist's impression, is a contact binary – two stars that orbit so close to each other that they are distorted by their gravity and actually touch. As a result, gas spirals off them into a disc surrounding both stars. The gravitational forces between them are so strong that they are slowing down the stars' orbits by nearly ten seconds each year.

See also: Star distances 84 · Stars colours 88 · Stellar evolution 92 · Star birth 94 · Introducing variable stars 108

Eclipsing binaries and double-double stars

24 LYRA *The small constellation of Lyra represents the lyre used by the legendary musician Orpheus on his journey into the underworld. The group is one of the brightest in northern skies. It lies on the edge of the Milky Way, flanked by Cygnus and Hercules, and is rich in interesting objects. These include arguably the sky's most famous multiple star system, and one of its most-observed nebulae. The annual Lyrid meteor shower, which peaks around April 21, also has its radiant here.*

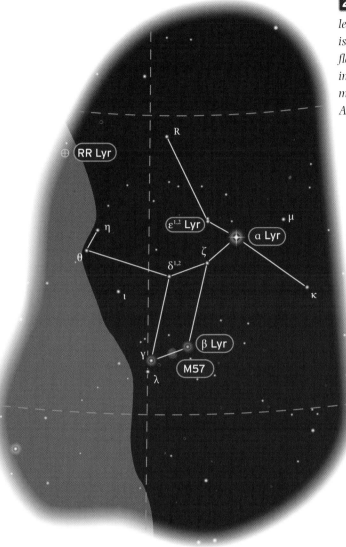

α Lyr Alpha Lyrae is better known as Vega, the fifth-brightest star in the sky at magnitude 0.0. Through a telescope, Vega appears to be a binary star, orbited by a magnitude 10.0, blue companion. This is merely a line-of-sight effect, at 25 light-years away, Vega is much closer than its blue neighbour. Vega is a white star, hotter and younger than the Sun, and is known to be surrounded by a ring of dust that may be forming planets.

RR Lyr RR Lyrae is the prototype of a class of variable stars that pulsate, changing their size and brightness in a regular cycle with periods of less than a day. RR Lyrae stars are often found in globular clusters, but this prototype lies in the main body of our galaxy, 745 light-years away. Its brightness varies from magnitude 7.0 to magnitude 8.1 in a 13.7-hour cycle.

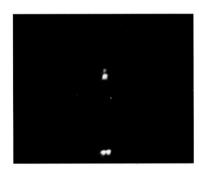

ε¹'² Lyr Epsilon Lyrae is the famous double-double – two pairs of stars in orbit around each other. The two pairs can be easily separated with binoculars, and shine at roughly equal brightnesses at magnitudes 6.0 and 5.2. Each of these stars can in turn be split in two by a small telescope with high magnification. The system lies 160 light-years away.

β Lyr Beta Lyrae is a multiple-star system, 880 light-years from Earth. Through a telescope it appears as a double star, with a bright, creamy coloured star and a magnitude 7.2 companion. But the cream-coloured star is a double too close to separate with even powerful telescopes. This star is an eclipsing binary, varying between magnitudes 3.3 and 4.4 in a 12.9-day cycle. Its stars are so close that they are pulled out of shape by each other's gravity, causing their light output to change as they rotate.

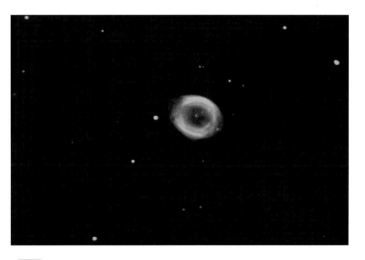

M57 This is the beautiful Ring Nebula, a planetary nebula created as a dying Sunlike star puffs off its outer layers into space. Through a small telescope, the nebula looks like a blue-green smoke ring shining at magnitude 9.7. The nebula lies 1,100 light-years away and is one of the brightest of its kind in the sky. The star at its centre shines at a very faint magnitude 15.0.

Introducing variable stars

Not all stars in the night sky shine with constant brightness. Many show marked changes in their apparent magnitude. These variable stars can arise not only in double-star systems, but also in a wide variety of single stars.

Over 50 different types of variable stars are known. Some change their brightness in a matter of hours, some over months and some over years. Some vary by just a fraction of a magnitude, whereas others change by many magnitudes. Furthermore, some have steady and predictable cycles, but others may lie dormant for years before suddenly brightening in a brilliant outburst.

Astronomers can tell a lot about a variable star from its light curve – a simple plot showing how its brightness varies over time. Other clues can come from the spectra of the stars, which reveal the presence of elements, and can also show Doppler shifts that indicate movement of the star's surface.

An eclipsing binary The stars are so close together that they appear to be one. Seen from Earth, each star passes in front of the other in its orbit and eclipses it to a greater or lesser extent.

Each type of variable is named after its prototype – the first star of that kind to be discovered. Two of the major classes (pulsating stars and explosions such as supernovae and novae) are discussed elsewhere. Other major classes include eclipsing, eruptive, and rotating variables.

Eclipsing binary variables, such as Algol in Perseus, are simple to understand. They are double systems in which one star periodically passes in front of and behind the other, so that the overall light reaching Earth from the system drops rapidly, then rises again. If the two stars in the system are of equal luminosity and size, then the two dips in brightness are equal, but more often the two stars are different, so the fluctuation in light will vary. If the two stars are close enough together, then gravity may distort one or both into an egg shape so that as they orbit, the visible surface area and therefore brightness, is constantly changing.

Erupting variables are stars that show violent activity in their upper layers. Our own Sun shows a pattern of increasingly violent surface activity over an 11-year cycle. However, the Sun's overall light output barely changes. By contrast, surface activity on an erupting variable has a far greater effect on the star's overall brightness. For example, flare stars are very dim, red-dwarf stars that nevertheless produce enormous flares, which can easily double their light output; and T Tauri variables are Sunlike stars going through a phase of rapid brightness changes before settling onto the main sequence. Other stars may be more brilliant than the Sun but still generate flares bright enough to affect their magnitude.

Sometimes a star changes brightness as it rotates. This may occur because a star in a close binary system has its shape distorted by its companion star. Alternatively, some giant stars, such as Betelgeuse, have a bright spot on one part of the surface where hot material is welling up from inside. Others – mostly faint dwarf stars – have enormous dark star spots a thousand times the size of the biggest sunspot.

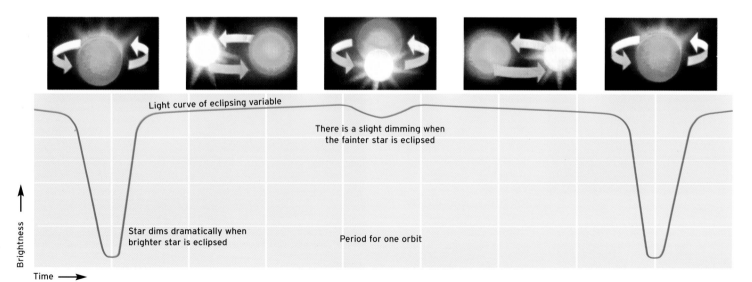

Light curve of eclipsing variable

There is a slight dimming when the fainter star is eclipsed

Star dims dramatically when brighter star is eclipsed

Period for one orbit

Brightness

Time ➝

See also: Star distances 84 · Multiple stars 106 · Pulsating stars 114 · Novae 124 · Open clusters 130

Algol—the winking demon

28 PERSEUS *The bright constellation Perseus lies on the Milky Way, bordered by Taurus to the south, and Cassiopeia and the faint circumpolar stars to the north. It is home to bright star clusters, nebulae and one of the most famous variable stars – the eclipsing binary Algol. The constellation represents a legendary Greek hero, the slayer of the gorgon Medusa and rescuer of the princess Andromeda from the sea monster Cetus. Algol, sometimes called the winking demon, is said to represent the eye of Medusa's head, whose gaze could turn the unwary to stone. Another variable in the constellation, Rho Persei, is a red giant which pulsates in size and brightness every 50 days. Perseus marks the direction of the closest spiral arm of our Milky Way Galaxy. It also contains the radiant of the Perseid meteor shower, one of the year's most reliable showers. This annual display every August is best seen by northern-hemisphere observers.*

NGC 869 **NGC 884** These two form the famous Double Cluster—a pair of bright neighboring star clusters 7,400 light-years away. With overall light equivalent to magnitude 4.3 and 4.4 stars respectively, the two clusters are both slightly larger than the Full Moon, and visible to the naked eye on a dark night. Each cluster contains around 200 hot, young stars, only a few million years old. They are best seen through binoculars, which will resolve many of their bright stars, and allow you to to appreciate the overall structure.

α Per Alpha Persei, also known as Algenib, is a yellow supergiant shining at magnitude 1.8. It lies in the midst of a broad scattering of faint stars called the Perseus OB-3 Association. This loose group, 550 light-years away, consists of relatively young and hot blue stars, drifting away from their common origin – an open cluster a few million years old.

M34 Messier 34 is another fine open cluster easily seen through binoculars. It contains around 60 stars shining at overall magnitude 5.8 and lies 1,400 light-years away. The stars in M34 are much older than those in the Double Cluster. They formed around 100 million years ago.

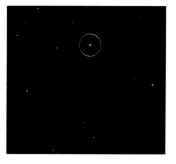

β Per Beta Persei, or Algol, is an eclipsing binary star. Lying 93 light-years from Earth, it consists of a brilliant white star and a fainter yellow-orange one, so close that they cannot be separated by even the most powerful telescopes. As the fainter star passes in front of the brighter one every 2.87 days, the brightness plunges from magnitude 2.1 (left) to magnitude 3.4 (right).

Clouds in space

The space between the stars may look empty, but within our galaxy it is crowded with clouds of gas and dust. Stars are born from these nebulae and return most of their mass to these clouds as they grow old and die.

The clouds between the stars, known as the interstellar medium, are composed of a variety of different materials, ranging from simple gases such as hydrogen and helium, to dark grains of heavier elements which astrophysicists term "metals". They even contain the complex carbon-based chemicals necessary for life. Clouds like this may linger for a 100 million years before a passing disturbance – perhaps a supernova, or the waves of pressure that regularly pass through a spiral galaxy's arms – triggers a burst of star formation. As the young stars form, they illuminate the gas around them, creating beautiful, multicoloured nebulae. Two varieties of bright nebulae are found—reflection nebulae, which simply reflect the light from nearby stars, and emission nebulae, which actually produce light themselves.

Reflection nebulae usually appear quite close to the stars that illuminate them. They frequently appear as hazy blue gas around young stars – as seen in clusters like the Pleiades – but can be almost any colour. However, interstellar clouds always appear bluer than the star that illuminates them because dust in the cloud scatters some of the star's blue light, making the star appear redder than it is. Astronomers are able determine the composition of the cloud by analysing its spectrum.

Emission nebulae are clouds of interstellar hydrogen gas that glow red when illuminated by ultraviolet radiation from youthful and hot, blue and white stars. Energy from the stars is absorbed by hydrogen atoms, boosting them into a state where they have lots of energy, but are unstable. They get rid of this energy by emitting low-energy red light. Many of the most famous nebulae in the sky, including the Orion and Carina Nebulae, glow in this way.

Although bright nebulae are some of the most spectacular and beautiful sights in the sky, they represent only a tiny fraction of all the interstellar clouds in our galaxy. The vast majority are not lit by either reflection or emission, and lurk in the blackness of space. Many are faintly warmed by the heat of distant stars, and glow dimly in low-energy radio or infrared radiation. Dark nebulae can be seen in visible light only where they are silhouetted against brighter backgrounds such as other nebulae. Orion's Horsehead Nebula is the most famous dark nebula, but several others are much easier to see. These include the Cygnus Rift and the Coal Sack, both of which appear as dark holes in the Milky Way.

The Horsehead Nebula Silhouetted against the field of a glowing emission nebula in the constellation of Orion, looms the Horsehead Nebula. The distinctively shaped horse's head is just part of a much larger dark nebula and can be seen only with large telescopes.

See also: Star distances 84 · Stellar evolution 92 · Star birth 94 · Planetary nebulae 116 · Supernovae 120

Darkness in the Southern Cross

α Cru Forming the southernmost point of the cross, Acrux is Crux's brightest star, shining blue-white at magnitude 0.8. It is 320 light-years from Earth. A small telescope reveals that it is in fact a double star, with twin blue-white components of magnitudes 1.3 and 1.7.

53 CRUX *Also known as Crux Australis—the Southern Cross—Crux is the sky's smallest constellation. It is, however, one of the most brilliant as it lies in one of the brightest areas of the Milky Way, between Centaurus and Carina. Crux takes its name from four bright stars at its centre, which form an easily discernible cross pattern whose long axis points down to the sky's south pole. The great dark nebula known as the Coal Sack obscures all but a few stars in Crux's southwest quadrant.*

β Cru Lying 350 light-years away, Beta Crucis, or Becrux, is a blue-white magnitude 1.3 variable star. Its small fluctuations are undetectable to the naked eye.

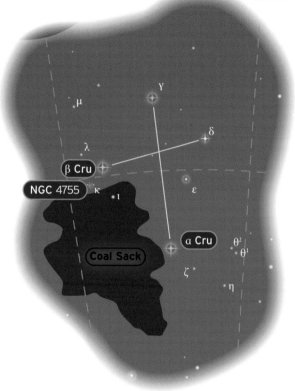

NGC 4755 Named the Jewel Box by the 19th-century astronomer John Herschel, NGC 4755 is a beautiful star cluster that lies around 7,600 light-years away. To the naked eye, it appears to be a magnitude 4.0 star, but binoculars reveal that it is actually made up of several dozen stars centred around the blue supergiant Kappa Crucis. Although most of its stars are also brilliant blue or white, an 8th-magnitude red supergiant lying near Kappa Crucis provides a startling contrast.

Coal Sack The Coal Sack is the sky's most prominent dark nebula, covering an area of sky 5° by 7° across. Lying roughly 400 light-years away, it blocks out a large area of the Milky Way.

OTHER FAMOUS NEBULAE

NAME	CONSTELLATION	DISTANCE	TYPE	AREA
Carina Nebula (NGC 3372)	Carina	9,000 light-years	Emission nebula	2° x 2°
Crab Nebula (M1)	Taurus	6,500 light-years	Supernova remnant	8 min x 6 min
Cygnus Rift	Cygnus	2,300 light-years	Dark nebula	–
Lagoon Nebula (M8)	Sagittarius	5,200 light-years	Emission nebula	1° x 30 min
Orion Nebula (M42)	Orion	1,500 light-years	Emission nebula	1.5° x 1°
Tarantula Nebula (NGC 2070)	Dorado	150,000 light-years	Emission nebula	30 min x 30 min
Trifid Nebula (M20)	Sagittarius	5,200 light-years	Emission nebula	30 min x 30 min
Veil Nebula (NGC 6992/6995)	Cygnus	4,200 light-years	Supernova remnant	1.5° x 1.5°

112 Red giants

After a star has exhausted the hydrogen fuel at its core, it starts to plunder other sources of nuclear fuel to keep shining. This activates rapid changes, and the star swells to hundreds of times its original size, becoming a blazingly bright red giant.

The nuclear fusion reactions that generate energy during a star's main-sequence lifetime gradually replace hydrogen in the star's core with helium, the next heavier element. Since the temperatures and pressures in the core are not high enough to burn helium, a star falters when the hydrogen in its core is used up, and it begins to burn hydrogen from the shell outside its core.

The switch to consuming hydrogen in a shell generates much more energy than when fusion was confined to the core, and the star's luminosity increases rapidly. At the same time, the extra energy heats up the star's interior, causing it to expand to many times its previous size. A star like the Sun can blow up to a hundred times its previous diameter – almost large enough to reach Earth's orbit – whereas more massive supergiants can grow 10 times bigger. Despite the increase in luminosity, the dramatic expansion cools the exterior of the star to around 3,000 °C (5,400 °F), causing it to turn red.

While the star's outer layers are expanding, the core is robbed of the energy source that supported it against its own gravity, and it begins to collapse. Over a few hundred million years, this collapse increases the temperature in the core of a star the size of the Sun to around 100,000 °C (180,000 °F). The searing heat produced enables the star to fuse helium nuclei into heavier elements – mostly carbon and oxygen – giving the star a powerful new energy source in its core. This boost in energy causes the star to expand, but the resulting drop in pressure disrupts the hydrogen-burning shell, so the star shrinks back and grows dimmer.

Helium burning in a star's core lasts only a fifth as long as the star's hydrogen-burning lifetime – two billion years for our Sun. As the helium in the core runs out, the star begins to burn helium in the shell, and once again the star swells in a second red-giant phase. For a star like the Sun, this is nearly the end of the story. Although its core collapses again, it can never reach the density needed to ignite the fusion of carbon or oxygen. For more massive stars, the process repeats itself several more times, as they burn heavier and heavier elements.

During the red-giant phase, convection currents carry material from the core to the surface in events called dredge-ups. The star's intense radiation blows much of this material away from the outer layers in strong stellar winds, and the entire star can frequently become unstable, pulsating regularly as a variable star.

Cool, red outer surface

Hydrogen

Convection current

Contracting core

Hydrogen-burning shell

Sooty exterior of heavy elements dredged up by convection currents

The phases of a red giant After the supply of hydrogen in the core of a star is exhausted, a process of expansion followed by contraction enables a star to use progressively heavier elements as fuel. As each new fuel is consumed, convection currents carry materials generated by nuclear fusion from the core to the surface.

See also: Star colours 88 · Plotting the lives of stars 90 · Stellar evolution 92 · Pulsating stars 114 · Planetary nebulae 116

Giants in Scorpius

79 SCORPIUS *The constellation Scorpius – the Scorpion – lies at the intersection of the ecliptic and the Milky Way close to the great star clouds of Sagittarius and the galactic centre. This position, in line with the centre of the galaxy, means that Scorpius is rich in star clusters and other galactic objects, including several globular clusters to the north of the Milky Way. Unlike many constellations, Scorpius actually resembles the object it is named after. The bright red star Antares marks the Scorpion's heart, and a distinctive curve of stars represents its tail. In ancient times, the neighbouring constellation of Libra was named Chelae Scorpionis, the Scorpion's Claws.*

α Sco Alpha Scorpii, also known as Antares – Rival of Mars – is one of the sky's brightest red giants. This supergiant, located roughly 600 light-years from Earth, has a luminosity 11,000 times brighter than the Sun. Antares contains the same mass of material as 15 Suns, and has swollen to over 700 times the diameter of the Sun.

β Sco Beta Scorpii is a line-of-sight double star at 530 and 1,130 light-years away, with blue-white components of magnitudes 2.6 and 4.9. These two stars can easily be separated in a small telescope.

M6 M7 Messier objects 6 and 7 are two of the sky's best open star clusters – groups of stars born from the same nebula at around the same time. M7 lies around 800 light-years from Earth and shines at a total magnitude of 3.3. It contains 130 stars in an area nearly one degree across, hence it is best seen in binoculars or telescopes at low magnification. M6 contains roughly the same number of stars, but is 2,000 light-years away. It shines at a total magnitude of 4.6 and has a distinctive butterfly-shaped pattern. The brightest star in M6 is the orange variable star BM Scorpii, which shines between magnitude 6.0 and 8.1.

ν Sco Nu Scorpii is a blue-white quadruple star system 440 light-years away. Binoculars can split it into two components of magnitude 4.0 and 6.3, and a medium-sized telescope will reveal that both of the components of Nu Scorpii are double stars.

M4 The globular cluster M4, is a group of 10,000 stars shining with a total magnitude of 7.4, 7,000 light-years from Earth. It has a comparatively loose structure, which allows even small telescopes to resolve the stars that lie at its outer edges.

114 Pulsating stars

As a star evolves into its giant phase, it often goes through a period of instability, where it changes its brightness and size in a regular cycle. These pulsating variables account for over two-thirds of all known variable stars.

Pulsating variables are stars that vary in brightness with predictable regularity. They are very widespread and important to our understanding of the Universe. They can be identified by their distinctive light curves, which show steady and regular changes – pulses – in the star's magnitude. The period of each pulse can range from a few hours to several years.

Pulsating stars are classified by their pulsation period and mass. Cepheid variables – after Delta Cephei – are yellow supergiants with masses greater than three Suns, pulsating in periods of a few days. RR Lyrae stars are small 0.3–0.7 solar-mass stars with periods of a few hours, and Mira stars are red giants with periods of hundreds of days.

When the light from any of these stars is split into a spectrum, its dark absorption lines are seen to be shifting regularly toward the red and then back to the blue end of the spectrum. Redshift and blueshift are caused when an objects recedes away and advances toward us – the Doppler effect. If the star is not orbiting a companion star, this effect must be caused by its surface physically pulsing in and out.

As a star's surface falls inward its brightness increases, and as it pushes outward its brightness decreases. It is at its brightest when smallest, and at its dimmest when largest. This can lead to distinct changes in colour, as the star's surface will be hotter and bluer when small and bright and cooler and redder when large and faint.

Since the star's overall brightness is changing, the amount of energy reaching its surface must be varying for some reason. As the nuclear reactions within the star cannot fluctuate, the cause of these changes must lie in something blocking the radiation on its way from the star's core to the surface.

The fluctuations may be triggered by a layer of helium around the star's core. The helium atoms have already had one of their two electrons stripped away by the heat inside the star. When the interior heats up as the star swells to a giant, the temperature eventually reaches a point at which the second electron can be stripped off. But in the process, the helium absorbs energy, becoming opaque and temporarily cutting off some of the radiation reaching the surface.

The drop in outward pressure allows the star's outer layers to fall inward. This compresses the star and heats it up until the pressure of radiation from inside becomes enough to push the upper layers out again. But the momentum of the expansion continues past the point of balance or equilibrium, so the pressure drops, the layers fall in again and the cycle is repeated. As the temperature and pressure near the star's core fluctuate, the thickness of the energy-absorbing helium layer also varies, causing the star's overall output of light to change.

Unstable stars Since variable stars change in luminosity, they move on the Hertzsprung-Russell diagram. Regions called instability strips mark the areas that pulsating stars must cross.

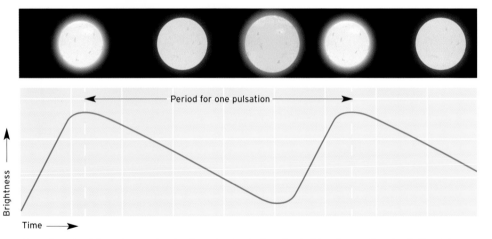

Cosmic rules Many types of pulsating star, including Cepheids, obey a direct relationship between the period of their pulsations, and their average brightness. By measuring the period precisely, the true brightness - luminosity - of a Cepheid variable can be found, and from that, its distance away from Earth can be calculated.

See also: Plotting the lives of stars 90 · Sunlike stars 98 · Introducing variable stars 108 · Red giants 112

The home of prototype variables

11 **CEPHEUS** *A comparatively faint group of stars, Cepheus lies between the bright star field of Cygnus and the north celestial pole. Even though it lies on the edge of the Milky Way, Cepheus does not have any bright star clusters. However, it does contain a wide range of double and variable stars, including the prototypes for three major classes of pulsating star – the Beta, Delta and Mu Cepheids. The constellation is named after a king from Greek mythology, the husband of Cassiopeia and father of Andromeda.*

δ Cep Delta Cephei is the prototype for the type of stars called Cepheid variables. It is a yellow supergiant, pulsating between 40 and 46 times the Sun's diameter in a period of 5 days, 9 hours. The star varies between magnitude 3.5 and 4.4. To estimate its magnitude, try comparing it to the nearby stars Gamma (magnitude 3.2) and Xi (magnitude 4.4).

β Cep The blue giant star Beta Cephei shines at magnitude 3.2 and is the prototype for a class of short-period pulsating stars that show very small and rapid changes in their brightness. Beta Cephei's brightness varies by about 0.1 magnitude in a cycle that repeats every 4.6 hours.

α Cep Alpha Cephei is one of the constellation's few bright stars that is not a variable. It is a white star, 59 light-years from Earth, shining at magnitude 2.4.

μ Cep Mu Cephei, also called the Garnet Star, is the reddest naked-eye star in the northern sky. It is another pulsating variable star, ranging from magnitude 3.4 to 5.1 in a complex 730-day cycle, and is the prototype for a group of long-period variables.

NGC 40 As well as variables, Cepheus also boasts a fine planetary nebula that is visible in small telescopes. NGC 40 is 2,000 light-years from Earth and shines at magnitude 10.7. It has a fairly bright central star of magnitude 11.6. Small instruments show the nebula as a compact but fuzzy, blue-green star.

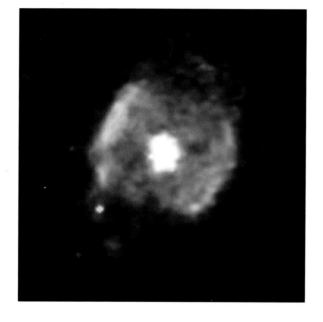

OTHER PULSATING VARIABLES

NAME	CONSTELLATION	MAXIMUM	MINIMUM	PERIOD	TYPE
Betelgeuse	Orion	0.2	1.2	Long, irregular	Red supergiant
Mira	Cetus	2.0	10.1	332 days	Long period red
R Coronae Borealis	Corona Borealis	5.8	15.0	Long, irregular	Carbon-rich star
RR Lyrae	Lyra	7.0	8.1	13 hours 36 mins	Cepheid relative
T Tauri	Taurus	9.3	13.5	Rapid, irregular	Young star
W Virginis	Virgo	9.5	10.8	17.3 days	Ancient Cepheid

116 Planetary nebulae

The death throes of Sunlike stars are revealed in delicate shells of glowing gas known as planetary nebulae. Among the most beautiful sights in the Universe, some of these nebulae form elegant rings, whereas many others display complex patterns.

The name planetary nebula was coined by the 18th-century astronomer and musician William Herschel, who commented on the similarity of certain nebulae to the ghostly, green discs of the planets. The term is misleading since, in reality, these clouds of gas have nothing to do with planets.

In fact, all planetary nebulae have dying stars at their centres. When a star with approximately the same mass as the Sun runs through its stock of hydrogen and helium fuel in its core, it cannot generate enough heat to burn heavier elements and begins to use gas in the layers outside the core as fuel for nuclear fusion. This marks the start of the star's growth into a red giant. The outward pressure of radiation generated by fusion reactions outside the core tends to blow the material in the outer layers of the star out into space, where it forms clouds. The planetary nebula is created when the red giant at the end of its life finally shugs off its outer layers. The removal of the star's outer layers reduces the gravitational pressure on the still-burning regions of the star, which lose heat and contract.

Despite the dwindling of the nuclear reactions, the outer layers of the star remain very hot – around 100,000 °C (180,000 °F). This heat radiates into space, illuminating the planetary nebula with intense radiation that causes the elements within it, such as hydrogen, helium, oxygen and nitrogen, to glow.

Different elements produce light at specific frequencies, producing a range of colours that reveals the composition of the cloud to observers. Astronomers are also able to measure the Doppler shifts of these signature wavelengths, and have discovered that the gas is usually expanding at speeds of about 32 km/s (20 mps). At this rate, most nebulae will disperse totally in a few tens of thousands of years.

Planetary nebulae exhibit a bewildering range of different shapes and structures. The simplest are perfectly spherical bubbles, which may appear from Earth as ring shapes (the circumference of the ring appears brightest because we are looking through a deeper layer of material at the edges of the bubble). But most nebulae are far more complex, with multiple overlapping rings of different colours and materials, large clumps of material that look like giant comets, and even pinched hourglass shapes.

Many of these structures are thought to arise when the tenuous, fast-moving shells of gas in the planetary nebula catch up and overtake denser gas thrown out earlier in the star's life. It seems that many red giants somehow throw off equatorial rings of gas which pinch the nebula in the middle, forcing it to blow out into huge lobes above the star's poles. Other strange and beautiful structures may be the work of a small companion star orbiting close to the red giant.

Planetary nebula Menzel 3 This image of the so-called Ant Nebula, taken by the Hubble Space Telescope, shows the fiery, asymmetrical lobes of gas around a dying star. The asymmetry may be caused by the pull of an orbiting companion star.

See also: Plotting the lives of stars 90 · Stellar evolution 92 · Sunlike stars 98 · Red giants 112 · White dwarfs 118

The flying horse and the fox

M15 The globular cluster M15 lies on the limit of naked-eye visibility at a combined magnitude 6.2. Through binoculars it appears as a fuzzy disc, but a small telescope will start to resolve stars around its edges. M15 lies 34,000 light-years away. It has an unusually bright core, and emits X-rays—both uncommon features that may point to a giant black hole at its heart, swallowing stars that stray too close.

27 36 PEGASUS & VULPECULA *Pegasus, the Winged Horse, is a large constellation best seen around October. The four brightest stars form a huge and distinctive square – officially the star located in the northeastern corner, Delta Pegasi, is now part of neighbouring Andromeda. Apart from the so-called Great Square, the region of sky around Pegasus is surprisingly empty. To its west lies the small and faint constellation of Vulpecula, the Fox, which holds the sky's most easily seen planetary nebula.*

β Peg The red-giant star Beta Pegasi, or Scheat, lies about 200 light-years away, and is an unpredictable variable, pulsating between magnitudes 2.3 and 2.7. At its brightest, it outshines Alpha Pegasi; at its faintest, it is similar to Gamma Pegasi.

NGC 7331 The barred-spiral galaxy NGC 7331 is the brightest galaxy in this region of sky. Shining at magnitude 9.5, it lies nearly edge-on to Earth, and appears as a small oval smudge through a small telescope. Larger instruments reveal its similarities to the nearby Andromeda Galaxy, although NGC 7331 is far more distant at 40 million light-years from Earth.

M27 The Dumbbell Nebula M27 is the most easily seen planetary nebula, shining at magnitude 7.6. It is visible through binoculars, and the bright central regions cover an area about a quarter the diameter of the Full Moon. Several stars lie within the nebula, but most are simply chance alignments. Its true source, a 13th-magnitude star on its way to becoming a white dwarf, lies approximately 1,000 light years away.

118 White dwarfs

What is left behind when a star dies? The last stages in any star's evolution are dependent on its mass, but the most common ending is as a white dwarf – a dense ball of hot gas, still glowing from the heat it once generated.

In the final stages of a Sunlike star's life, it rids itself of a shell of gas from its outer layers, which forms a planetary nebula. At the centre of the nebula, the remnant of the star stops its nuclear fusion reactions as temperatures and pressures fall too low. For the first time since it was born, the star has no outward surge of radiation to counteract the inward pull of its gravity, and it begins to fall inward.

The collapse into a white dwarf is a slow process, as the surviving inner regions of the star are still extremely hot and have a natural tendency to expand. But the gravity of the core, which may still contain a solar mass of material, is too strong to resist.

A mature white dwarf whose planetary nebula has long since disappeared has normally condensed to about the size of Earth, and has a surface temperature of around 100,000 °C (180,000 °F). But it shrinks no further than this. The high temperature within ensures that its atoms remain separated into atomic nuclei and negatively charged electrons

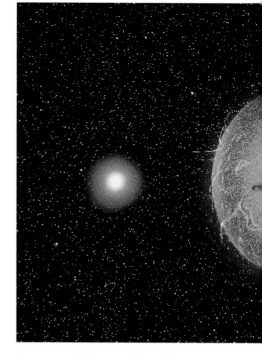

Extreme neighbours This artist's impression shows a small, hot and comparatively dim white dwarf orbiting with an enormous, cool but brilliant red giant.

and, when the star reaches a certain size and density, repulsion between electrons prevents it from shrinking further. This repulsion remains constant even as the white dwarf slowly cools to a theoretical (but never observed) black dwarf. By this point, the star's density is such that a teaspoon of it would weigh a ton.

One odd aspect of white dwarfs is that the more massive they are, the smaller their diameter. The largest white dwarfs may be the size of a gas-giant planet like Uranus, but the smallest, heaviest ones may be smaller than Earth. This is because additional mass makes for stronger gravity, which packs the star's material more tightly.

Eventually, however, a limit is reached. If a white dwarf contains more than 1.4 solar masses of material (called the Chandrasekhar limit) its gravity will overwhelm the outward pressure of electrons. Such an object will shrink to a much smaller size, crushing all its matter together and forming a fantastically dense object called a neutron star.

A dying star Even the closest white dwarfs are extremely faint objects. They mostly reveal their presence through their pull on other stars, which causes them to wobble on their path through space. Such wobbles were the first clues that white dwarfs even existed. Sirius B, shown here with Sirius, was the first to be observed directly and was not seen until 1862.

See also: Stellar evolution 92 · Sunlike stars 98 · Red & brown dwarfs 102 · Multiple stars 106 · Planetary nebulae 116

The river of the heavens

55 **ERIDANUS** *The long and winding constellation of Eridanus, the River, stretches from just south of the celestial equator near Orion, into the far south of the sky. Its stars are mostly faint and undistinguished, but its position in the sky, well away from the Milky Way, gives a clear view to several distant galaxies in deep space. Eridanus also contains the brightest and most easily seen white dwarf in the sky, and one of the nearest Sunlike stars to Earth.*

NGC 1535 This small but distinctive planetary nebula shines at magnitude 9.6. Although large telescopes and high magnifications are needed to show the shape of its disc, it is easily identified against the background stars because of its unusual blue-grey colour. NGC 1535 lies about 6,850 light-years from Earth.

NGC 1337 Although NGC 1337's magnitude is only 12.6, because this spiral galaxy lies edge-on to Earth, its light is concentrated in a small region of the sky, making it visible through small telescopes.

ε Eri The yellow star Epsilon Eridani, magnitude 3.7, lies just 10.5 light-years away and is one of the most Sunlike of nearby stars. It is slightly dimmer and more orange than the Sun, but it is still a potential haven for life. So far, astronomers have not found any planets in orbit around the star, which is only a few hundred million years old. But they have photographed a dust ring around it, which could indicate that planets are forming.

o² Eri The triple star system Omicron² Eridani (also known as 40 Eridani) is one of the closest star systems to our own, and home to an easily seen white dwarf. A small telescope reveals that the naked-eye yellow star, magnitude 4.4, has a white companion star of magnitude 9.5 close to it. This white dwarf, 16 light-years away, has a close companion of its own – a dim red-dwarf star of 11th magnitude, only visible with a larger telescope.

NGC 1300 This galaxy near the centre of Eridanus shines at magnitude 11.1 and is a good example of a face-on, barred-spiral galaxy. Small telescopes will show the galaxy's central bar as an elongated patch of light with a bright centre, but larger instruments are needed to see the spiral arms trailing away from the ends of the bar.

STAR STORIES

Grand Theft Chariot

Eridanus is one of the sky's oldest constellations, dating to Ptolemy's original list of 48 compiled in the first century. It has been associated with many different rivers in its time, including the Egyptian Nile, and the Italian Po. In mythology, however, it represents the river where the god Phaethon fell to Earth after stealing the chariot of the Sun God Helios and driving it recklessly across the sky. He was stopped when Zeus struck him down with a thunderbolt.

α Eri Alpha Eridani, or Achernar (River's End), is the constellation's southernmost and brightest star. It is a magnitude 0.5 blue-white giant, 91 light-years from Earth.

Supernovae

The most massive stars die spectacularly: instead of dwindling slowly, they experience runaway nuclear reactions that culminate in a vast explosion called a supernova. During this brief flash of brilliance, they can release more energy than an entire galaxy.

A Type II supernova marks the end of the evolution of a high-mass star. Another kind of supernova is also known and referred to by astronomers as Type I supernova, which is a special type of nova. Here, we will concentrate only on the deaths of massive stars.

A Sunlike star uses hydrogen as fuel for the nuclear fusion reactions at its core, before moving on to consume the helium generated by the fusion of hydrogen. At the end of each phase, the star swells into a red giant. The second red-giant stage marks the end of its life, since its core can never reach sufficient temperature or pressure to burn the products of helium fusion.

In stars that weigh eight solar masses or more, a third fusion phase can happen due to the enormous weight of the star's outer layers. The core's temperature and pressure reach a point where carbon and oxygen (the major products of helium fusion) begin to burn. This is the beginning of a process where the star makes heavier and heavier elements, but with rapidly dwindling energy efficiency.

The central regions of such high-mass stars become factories for the production of heavy elements. Carbon burning produces oxygen, neon, sodium and magnesium, which in turn ignite to create a range of elements, from sulphur up to iron.

Each new element is used up more quickly than the last, releases less energy, and creates less of the next element. The core becomes smaller and denser as each fuel supply is consumed, and then disaster strikes: iron is the first element that takes more energy to create than it releases during fusion. The core's

power supply is suddenly cut off, and it collapses under its own weight, freezing into an extremely dense, solid sphere, just several tens of kilometres across. At the same time, the iron in the core becomes unstable, disintegrating into helium nuclei, and then into neutrons.

Meanwhile, robbed of support, the layers around the core plunge down at speeds of thousands of kilometres per second to fill the space left by the reduced core. When they hit the rigid remnant of the core, they rebound with a force that rips the star apart, releasing enormous amounts of energy in an outburst that lasts for months. As the star tears itself to pieces, material in the outer layers is bombarded with neutrons from the core, generating elements heavier than iron.

Death throes of a massive star The core of a massive star is like an onion, with the lightest elements at its edges and the heaviest elements at its heart. The star's atmosphere is supported by the force of the radiation from the core but, with the fusion of iron, the enormous pressure collapses the core and triggers a supernova.

Outward force of radiation

Atmosphere of star near the core

Inward pressure from star's outer layers

Core

Energy source cuts out

Collapse of core

Southern supernova A supernova lasts for several months and outshines all surrounding stars. SN 1987A in the Large Magellanic Cloud increased in brightness a millionfold.

See also: Stellar evolution 92 · Stellar heavyweights 104 · Neutron stars & pulsars 122 · Novae 124 · Black holes 126

The raging bull

M1 Caught between the bull's horns is the Crab Nebula, M1 – the glowing remains of a supernova 6,500 light-years away, was seen by Chinese and Native American astronomers in A.D. 1054. Glowing at magnitude 8.4, the Crab appears as an oval wisp of light through small telescopes. Much larger instruments and long photographic exposures are needed to reveal its shredded, filamentary structure.

32 **TAURUS** *Clearly resembling the front of a charging bull, Taurus is one of the sky's most distinctive constellations. The red-giant star Aldebaran marks its baleful eye, Hyades defines its face, and the Pleiades forms its shoulder. Taurus faces the hunter, Orion, and is best seen in evenings around January. Bordering on the Milky Way, it contains some of the sky's best star clusters, as well as the Crab Nebula supernova remnant. However, the constellation is too close to our galaxy's plane to offer views of other galaxies.*

M45 The Pleiades star cluster M45, also known as the Seven Sisters, is probably the most beautiful star cluster in the sky. Covering an area more than twice the size of the Full Moon, to the naked eye it has seven stars in a distinctive shape. Viewing the cluster through binoculars reveals dozens more hot blue-white stars – the cluster contains more than 100 in all. The Pleiades lies about 400 light-years from Earth, and contains some of the youngest stars in the sky. They are thought to have formed about 50 million years ago, and are still surrounded by a faint reflection nebula visible in large telescopes.

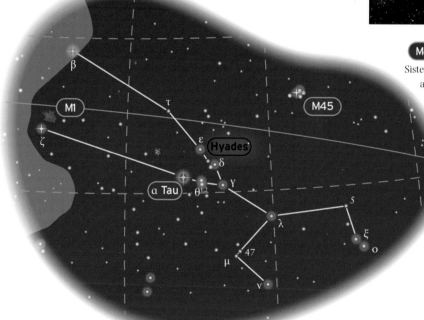

α Tau The red giant Aldebaran lies in the midst of the Hyades star cluster, but is actually much closer to Earth, about 65 light-years away. Aldebaran is an irregular variable star, pulsating between magnitudes 0.8 and 1.0 with an unpredictable period. It is also a double star, with a dim red dwarf of magnitude 13.4 lying nearby.

Hyades The V-shaped Hyades cluster, lying 160 light-years away, beyond Aldebaran, is larger than the Pleiades, scattered over a much broader region of sky, and contains more members. In fact, the Hyades is so big that it is best seen through binoculars, as its overall shape tends to be lost in a telescope's smaller field of view.

122 Neutron stars and pulsars

The cataclysmic supernova that destroys a massive star can give birth to a new and strange object. The unstoppable gravity of the star's collapsing core forges a superdense rigid sphere, packed with the same incredible density as an atomic nucleus.

Neutron stars are similar to white dwarfs in that both are collapsed cores of dead stars. But neutron stars are far smaller than white dwarfs and can have masses of up to five Suns. It is this high mass that makes the collapse to a neutron star brief and violent. The star is in fact an incredibly dense mass of neutrons – neutral subatomic particles. In the extreme conditions of a collapsing giant star,

positively charged protons unite with negative electrons to form neutrons. Lacking an electrical charge, these can be packed to an incredible density.

One strange characteristic of neutron stars is that their diameter shrinks as their mass increases. Like white dwarfs, they also have an upper mass limit, above which the neutrons disintegrate under gravity, and the star collapses to a black hole. Neutron stars

have one of the weirdest structures in the Universe – a solid surface that encases the seething hot material inside, with an inch-thick atmosphere of hot plasma trapped above it. They spin in fractions of a second, and have intense magnetic fields, trillions of times stronger than the Sun's.

It is these features of neutron stars that give away their presence. Although intensely hot, they are so small that they do not emit enough light to be seen normally. However, these strange objects, called pulsars and magnetars, can be detected through other emissions.

Pulsars are cosmic lighthouses, beacons that emit bursts of radio waves, visible light, and other radiation with precisely repeating periods ranging from a few milliseconds to more than a second. The first pulsar discovered was nicknamed Little Green Man 1, since astronomers at first thought it might be a signal from an alien civilization.

A typical pulsar is a rapidly spinning neutron star with a magnetic field tilted at an angle to its axis of rotation. The star's intense magnetism sweeps up electrons surrounding it, sending them ricocheting back and forth and generating radio waves in just the same way as a radio antenna on Earth.

Magnetars are strange objects that emit sudden bursts of high-energy gamma radiation, which then dissipate in a series of diminishing pulses. Magnetars are thought to be pulsars with even more intense magnetic fields – up to a thousand times stronger than a normal neutron star. As the field sweeps around the star, it interacts with the plasma atmosphere, slowing the magnetar's spin, and setting up powerful waves. These disrupt the star's crystalline surface, causing starquakes that allow hot material to burst through the crust and create a shining beacon of intense gamma rays.

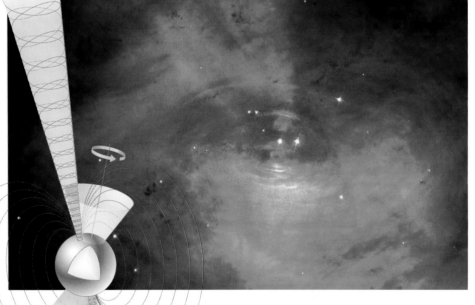

Remnants of a great supernova This picture taken by the Hubble Space Telescope peers deep within the Crab Nebula in the constellation Taurus. Its central star is a pulsar and flashes 30 times per second, emitting radio waves, visible light and X-rays. The explosion that produced the pulsar occurred more than 900 years ago.

Cosmic lighthouses A pulsar's signal is created directly above its magnetic poles. If the magnetic and rotation axes of the star are tilted to one another, then rotation causes the beam of radiation to sweep around in an arc that may point toward Earth once in every cycle.

See also: Stellar evolution 92 · White dwarfs 118 · Supernovae 120 · Black holes 126 · The invisible Universe 182

Finding the pulse of Vela

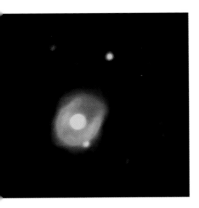

NGC 3132 The planetary nebula NGC 3132 is one of the brightest such objects in the sky, shining at magnitude 8.2, and with a diameter larger than Jupiter's as seen from Earth. It lies about 1,800 light-years away and has an unusually bright central star, shining at magnitude 10.1, which is visible through a small telescope.

86 VELA *The constellation of Vela, the Sail, once belonged to the much larger constellation of Argo Navis, the ship on which Jason and the Argonauts sailed in search of the Golden Fleece. It is still one of the southern sky's finest constellations, best seen in the evenings around March and April. Vela lies in a rich region of the Milky Way and hides one of the few pulsars to flash in visible light, as well as two supernova remnants. The Vela pulsar is one of the most intense in the sky, shining at visible as well as radio wavelengths. The pulsar flashes once every 89 milliseconds, and is gradually slowing down, at a rate of 10.7 nanoseconds (billionths of a second) every day.*

NGC 3201 The globular cluster NGC 3201 lies 17,000 light-years away from Earth and covers an area of sky half the size of the Full Moon. Shining with an overall magnitude of 6.8, it has a comparatively loose structure, and small telescopes will show stars nearly all the way to its centre.

γ Vel Vela's brightest star is Gamma Velorum, about 840 light-years away. Shining at magnitude 1.8, it is the brightest Wolf-Rayet star – a supermassive star that is losing large amounts of mass through its strong stellar winds. It is also a spectacular double star, with a magnitude 4.2 companion.

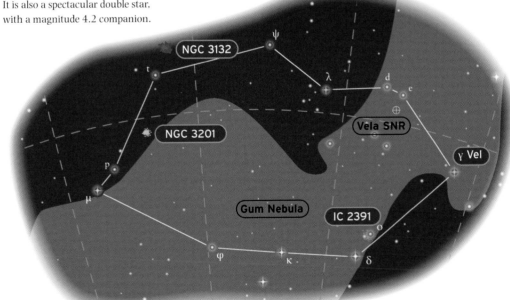

IC 2391 The large, bright open cluster IC 2391 shines at magnitude 2.6 and covers an area twice the diameter of the Full Moon, so it is best seen with binoculars. It is so prominent that it has been nicknamed the Southern Pleiades. The cluster lies about 450 light-years away and contains about 30 stars.

Vela SNR The Vela Supernova Remnant is a faint area of shredded gas, only visible by long-exposure photography. At its heart lies the Vela pulsar, remnant of a stellar explosion that brightened the skies of Earth 11,000 years ago. The pulsar flashes faintly 11 times a second and is a powerful X-ray source.

Gum Nebula Although only visible in long-exposure photographs, Vela's second supernova remnant, also known as the Gum Nebula, covers a huge area of sky – more than 60 degrees across. It is the aftermath of a supernova that exploded about 11,000 years ago, now expanded into a huge bubble more than 3,000 light-years across.

124 Novae

Novae are the most violent of variable stars: objects that flare up to hundreds of times their normal brightness, then fade over days or weeks. They are usually far less bright than supernovae, but sometimes a nova system can turn into a supernova.

The ancients thought that novae were new stars and named them after the Latin word for new. Novae only ever occur in spectroscopic binaries – double-star systems where the stars are too close to separate with a telescope, and can only be detected by Doppler shifts in their spectra.

The spectra of such stars also reveals the details of the system. Normally, the visible star is large and bright, nearing the end of its life and often swollen into a red giant. The other star, although small and faint, exerts a strong pull on its companion and carries the signature spectral lines of carbon, oxygen and other heavy elements, which show that it is a white dwarf.

Nova systems start out as close binaries where one star is more massive than the other, but not massive enough to destroy itself and its companion in a supernova. The more massive star evolves more quickly, swelling to a red giant, blowing off its outer layers and becoming a white dwarf while its companion is still on the main sequence. Eventually, the lower-mass star also begins to evolve and move off the main sequence. As it swells to its giant phase, its outer layers come close to the white dwarf's gravitational field. Material is then pulled away from the top of the giant's atmosphere, and spirals down onto the white dwarf. It seems that the nova explosions themselves take place on the surface of the white dwarf.

The gas ripped from the giant star is largely hydrogen, and this forms a highly compressed atmosphere around the white dwarf. Temperatures and pressures can grow to rival those inside a star. When the temperature rises above 10 million °C (18 million °F), the hydrogen begins to burn in a runaway nuclear fusion reaction, engulfing the star in a brilliant outburst of radiation. Something similar can also happen in certain neutron stars called bursters. These systems produce an intense burst of X-rays, rather than visible light, and the eruption only lasts for a few seconds.

In a few nova systems, when the white dwarf has about 1.4 times the mass of the Sun, it hovers on the edge of the Chandrasekhar limit – between a white dwarf and a neutron star. A small amount of mass transferred from the companion star can tip the white dwarf over the edge and trigger a Type I supernova. The internal temperature and pressure of the star rises to the point where fusion of carbon can suddenly begin, and a runaway nuclear reaction overtakes it, blowing it apart in a huge explosion.

This type of supernova is very useful for astronomers trying to measure the distance between galaxies. Because they always involve the same event – the destruction of a white dwarf with 1.4 solar masses of material – they always release the same amount of energy. So the brightness of this type of supernova in a distant galaxy is a good indicator of its distance, a useful standard candle.

Bonnie and Clyde binaries This artist's impression shows how gas is snatched away from a large red giant by its companion white dwarf. The mass-transfer process can continue for as long as the red giant remains within reach of the white dwarf, creating a recurrent nova with periodic explosions, or it can end in a blaze of glory–a Type I supernova.

See also: Stellar evolution 92 · Multiple stars 106 · Red giants 112 · Supernovae 120 · The expanding cosmos 166

The crown jewels

13 **16** **CORONA BOREALIS & DRACO** *The Northern Crown is a relatively faint, mid-northern constellation, bordered on either side by Boötes and Hercules. Draco, the Dragon, a little way to its north, is also faint and snakes halfway around the pole, coiling around the constellation Ursa Minor. Corona and Draco are ancient constellations, and Corona in particular resembles the crown it is named after. Because these constellations lie well away from the Milky Way, they offer good views of the depths of space. Corona, for example, contains a huge cluster of galaxies, although all are of 16th magnitude or less.*

M102 This object was confusingly catalogued by Messier, but most astronomers believe the number refers to a bright lenticular galaxy, also known as NGC 5866. This galaxy, shining at magnitude 10.8, and visible in small telescopes, is an example of a type of galaxy that seems to bridge the gaps between the young, active spirals and the giant balls of stars known as elliptical galaxies.

NGC 4236 The barred-spiral galaxy NGC 4236 is an attractive swirl of stars shining at magnitude 10.1. A small telescope will show the bright concentration of stars along its bar.

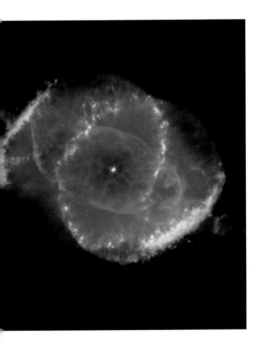

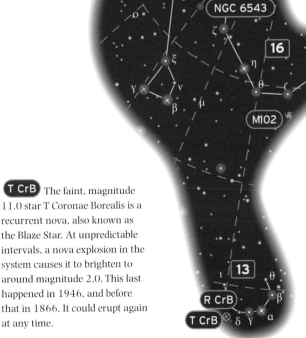

NGC 6543 The Cat's Eye Nebula, NGC 6543 is small but bright and shines at magnitude 8.8. At 2,000 light-years away, it is easily seen because of its high contrast against the dark background skies. High magnifications are needed to reveal the nebula's circular shape, but its blue-green colour helps to distinguish it from background stars at first glance.

T CrB The faint, magnitude 11.0 star T Coronae Borealis is a recurrent nova, also known as the Blaze Star. At unpredictable intervals, a nova explosion in the system causes it to brighten to around magnitude 2.0. This last happened in 1946, and before that in 1866. It could erupt again at any time.

R CrB The strange star R Coronae Borealis is a very unusual variable star that does just the opposite of a nova. Normally, this yellow supergiant shines at magnitude 5.8 but, every few years, its magnitude rapidly drops, sometimes as low as magnitude 15.0. This catastrophic dimming is thought to be caused when the carbon-rich star ejects huge shells of soot from its atmosphere into clouds that obscure most of its light for weeks at a time. It lies around 6,000 light-years from Earth.

126 Black holes

Black holes are one of the strangest phenomena the Universe has to offer: infinitely dense points of matter with such intense gravitational fields that, once within a fatal perimeter called the event horizon, not even light can escape them.

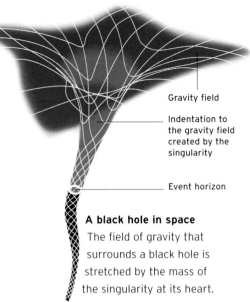

Gravity field

Indentation to the gravity field created by the singularity

Event horizon

A black hole in space
The field of gravity that surrounds a black hole is stretched by the mass of the singularity at its heart.

Black holes can form when a giant star with a core weighing more than about five solar masses destroys itself in a supernova. As the star's heart collapses, it rapidly shrinks past white-dwarf diameter, and the heavy elements within disintegrate into neutrons as it goes. If the core were lighter, the pressure between these neutrons would become strong enough to resist a total collapse, and the star's remains would stabilize as a city-sized neutron star. In this case, however, even the forces between neutrons cannot withstand the enormous inward pressure, and the star continues to collapse as the neutrons shatter into their component particles called quarks. A black hole forms when the gravity of the collapsing star grows so strong that the escape velocity (the speed at which an object must travel to get away from its gravitational pull) exceeds the speed of light— 300,000 km/s (186,000 mps). By contrast, the escape velocity at Earth's surface is a mere 11 km/s (7 mps). After the collapsing star passes this critical point, known as an event horizon, no further light is emitted and it is sealed off from the outside Universe. Inside the black hole, the star continues to collapse until it reaches a microscopically small, almost infinitely dense point called a singularity.

Although a black hole, by definition, cannot be seen directly, astronomers can identify their effects. For example, if a high-mass star in a binary system is orbiting around an unseen object of even higher mass, then that object is very possibly a black hole. More conclusive evidence is provided when a binary system like this also emits X-rays. This occurs when the black hole is close enough to the normal star for its gravity to rip matter away from the star's outer atmosphere. This material is drawn into a spiral accretion disc, where it is torn apart into atomic and subatomic particles, emitting X-rays, before disappearing into the hole itself.

Black holes can also be formed from giant clouds of material at the centre of galaxies. Such supermassive black holes have a mass millions of times that of the Sun. Some astronomers believe that there may also be a third, miniature type of black hole, scattered across interstellar space, left over from the creation of the Universe.

Fatal attraction As the blue giant star in this binary system orbits its companion black hole, its atmosphere is stripped away into an X-ray emitting accretion disc.

See also: Stellar heavyweights 104 · Multiple stars 106 · Supernovae 120 · The centre of our galaxy 136

Cygnus' black heart

14 CYGNUS *Representing a swan, Cygnus incorporates one of the richest regions of the Milky Way, containing more than its share of clusters, nebulae and double stars, as well as more exotic objects such as the black hole candidate Cygnus X-1. The cross shape made by Cygnus' brightest stars is easily identified on northern summer nights, when the swan's long neck extends south along the Milky Way, and its wings are outstretched across it. Because of this distinctive shape, Cygnus is also known as the Northern Cross.*

M39 The bright, open star cluster M39 shines at magnitude 5.3, and covers an area half the diameter of the Full Moon. It contains 28 or more stars of magnitude 7.0 and below, and lies about 880 light-years away.

α Cyg The tail of Cygnus is formed by Alpha Cygni, also known as Deneb, a magnitude 1.3, blue-white star. This incredibly luminous supergiant lies around 3,000 light-years from our Solar System.

NGC 7000 Called the North America Nebula because of its shape, NGC 7000 is just about visible with the naked eye in good conditions. This interstellar cloud is a complex mixture of emission, reflection and dark nebulae, about 1,500 light-years from Earth.

Cyg X-1 Cygnus X-1 is an intense source of X-rays, locked in orbit with a magnitude 8.9, blue supergiant. It is almost certainly a black hole.

NGC 6992 The Veil Nebula is the remnant of a supernova explosion that occurred around 30,000 years ago. It forms the brightest part of a complex structure called the Cygnus Loop, which is best seen in long-exposure photographs.

Cygnus Rift A dark hole in the Milky Way runs the whole length of Cygnus, separating a large cloud of stars from the rest. This hole is caused by a huge dark nebula called the Cygnus Rift, similar to the southern sky's Coal Sack.

β Cyg Located 390 light-years from Earth, Beta Cygni, or Albireo, is one of the sky's most famous double stars. Its two elements can easily be separated with binoculars and display beautiful complimentary colours. The brighter star is a yellow giant of magnitude 3.1, and its companion is blue-green, and shines at magnitude 4.7.

128 Strange stars

Sometimes astronomers encounter stars that don't obey any of the normal rules. Often two or more types of star combine to produce a hybrid that may exhibit strange variability or radiate at unusual wavelengths.

By their very nature, strange stars are one-of-a-kind objects. Explaining these systems when having only the angle of view afforded from Earth, and seeing just one stage in a complex evolutionary history, can tax astronomers to the limit.

For example, Epsilon Aurigae has strange eclipses every 27 years. Since the eclipses last two years, the object that causes them must be huge, and should block out the star completely. But instead, much of the star's light seems to shine through the object. The best current explanation is that Epsilon is orbited by a semitransparent disc of matter – possibly a planet-forming disc around the pair of faint dwarf stars which orbit Epsilon.

An even more bizarre object is SS 433 in the constellation Aquila. This star emits radiation over a huge range of frequencies, and Doppler shifts in its light seem to show that it is moving toward and away from us at high speed simultaneously. Astronomers think that the radiation must actually come from jets – one pointing toward us, and one away. The jets are probably caused by a dense neutron star or black hole pulling material away from a large, old star.

Polar variables are another strange type of system. They involve a rapidly spinning, dense white dwarf with a powerful magnetic field, orbiting a low-mass, red-dwarf star in just a few hours. The white dwarf's gravity pulls material away from the red dwarf, but the strong magnetic field prevents it from spiralling down onto the star as in normal novae systems. Instead, gas is swept up in the field and dumped directly onto the white dwarf's poles, generating intense radiation, which is polarized (all aligned in a single plane), and varies as the white dwarf spins.

Auriga mystery The bizarre variable Epsilon Aurigae may be a normal star periodically eclipsed by a tenuous protoplanetary disc, which is orbiting a close pair of faint stars.

See also: Stellar evolution 92 · Multiple stars 106 · White dwarfs 118 · Neutron stars & pulsars 122 · Black holes 126

The celestial charioteer

4 **AURIGA** *Known as the Charioteer, this bright constellation rides high in the sky on northern winter nights. Lying across the Milky Way, it is rich in star clusters and unusual stars, including the extremely strange Epsilon Aurigae. The brightest star, Capella, is associated with the she-goat Amaltheia, who nursed the young Zeus, king of the Greek gods. The small and distinctive triangle of stars just to the southeast represents Amaltheia's kids.*

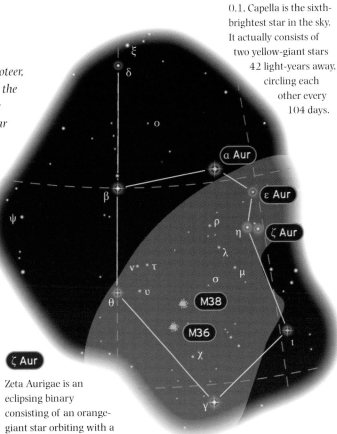

α Aur At magnitude 0.1, Capella is the sixth-brightest star in the sky. It actually consists of two yellow-giant stars 42 light-years away, circling each other every 104 days.

M38 M38 is an attractive open cluster for binocular observers. It contains 100 stars that shine at a combined magnitude of 6.8, and is slightly farther from Earth than M36 at 4,300 light-years. M38's brightest star is a yellow giant, some 900 times more luminous than the Sun.

ζ Aur Zeta Aurigae is an eclipsing binary consisting of an orange-giant star orbiting with a smaller blue companion.

ε Aur Epsilon Aurigae lies 2,000 light-years away. It normally shines at magnitude 2.9 but, at its next eclipse in 2009, it will drop to magnitude 3.8.

M36 The bright star cluster Messier 36, which is 4,000 light-years away, contains about 60 stars of magnitude 9 and below, whose combined brightness is magnitude 6.5. It is easily spotted with binoculars, and small telescopes will reveal its individual stars, grouped within a region roughly half the size of the Full Moon.

OTHER STRANGE VARIABLES

NAME	CONSTELLATION	FEATURES	EXPLANATION
Black Widow Pulsar	Sagitta	Variable period pulsar	Pulsar beam is blasting through a companion star
Cygnus X-1	Cygnus	X-ray and radio emitter	Black hole disrupting a companion supergiant star
PSR 1937+21	Vulpecula	Fastest-spinning pulsar	Pulsar spins up as it grabs material from companion
Scorpius X-1	Scorpius	Bright X-ray source	Material spiralling from a star onto a neutron star
SS 433	Aquila	Twin quasarlike jets	Material spiralling from a star into a black hole

130 Open clusters

The sky in and around the Milky Way is crowded with clusters of stars – stellar groups all born at the same time from the same interstellar cloud of dust and gas. Astronomers can learn a lot from these groups about the way stars interact and evolve.

There are two types of star cluster: open clusters and globular clusters. Open clusters are scattered and often contain young, hot stars, and even traces of the nebula that formed them.

Clusters form with a wide range of stars of different masses from blue giants to red dwarfs. Because all these stars were born within a few million years of each other, they can be reckoned to be the same age in cosmic terms, so astronomers can observe how some live their lives faster than others. The heaviest stars form most quickly and join the main sequence while smaller stars are still condensing. As these massive, hot, blue stars begin to

shine, their radiation blows away much of the nearby nebula, denying smaller nearby stars the chance to grow and often creating an emission nebula. As the nebula ages, smaller and lighter stars join the main sequence. But by the time the smallest dwarfs begin their lives, the massive blue giants are already moving off the main sequence, swelling into red giants as they exhaust their primary sources of fuel.

Because a cluster's stars lie roughly at the same distance, their apparent magnitude is closely related to their real luminosity. The concentration of stars in different parts of the main sequence can be used to work out the cluster's

age. Most clusters are relatively young (just a few hundred million years old) because the older a cluster gets, the more dispersed it becomes. Each star has its own motion, and they gradually scatter across space.

The movements of stars within a cluster also allow astronomers to work out its distance using a form of parallax effect. If the cluster is nearby, then its members will appear to be moving apart in the sky far more rapidly than the members of a more distant cluster. This moving-cluster method works at far greater distances than normal parallax, and distances to star clusters such as the Hyades in Taurus and Praesepe in Cancer now form the backbone of the cosmic distance scale.

Seven sisters The Pleiades in Taurus are about 100 young stars, still surrounded by shreds of the gas from which they formed.

Clusters in Cancer

7 **CANCER** *The zodiac constellation Cancer, the Crab, is faint and indistinct, although easily found as it lies between the much brighter Gemini and Leo. Best seen in the evenings around March, Cancer lies a little distance from the Milky Way, but is still too close to its plane to offer clear views of distant galaxies. However, the constellation does hold several rich star clusters, including the well-known Beehive Cluster. Cancer is connected mythologically with Hercules as the hero crushed the crab beneath his foot as he was performing one of his 12 tasks.*

α Cnc Alpha Cancri, known as Acubens – the Claw – is a double star consisting of a magnitude 4.2, white star accompanied by a small, 12th-magnitude companion only visible in medium-sized telescopes. It lies 175 light-years from Earth.

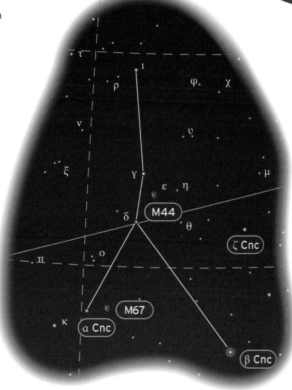

Field of view = 10'

β Cnc Owing to one of the many slips in Bayer's constellation catalogue, the brightest star in Cancer is actually designated Beta Cancri, not Alpha. Appearing at magnitude 3.5 from Earth, Beta (also known as Altarf) is an orange giant some 290 light-years away.

ζ Cnc Zeta Cancri is an attractive triple star only 84 light-years away, and is a good test of small telescopes. It consists of three yellow stars of magnitudes 5.2, 5.8 and 6.2, with the magnitude 6.2 star not visible at this scale. Small telescopes should be able to resolve the system into two stars with ease (left) but high magnification, a steady atmosphere, or a larger telescope is needed to separate the closer pair.

M44 Easily visible with the naked eye, the magnitude 3.9 star cluster Messier 44 is better known as the Beehive Cluster, or by the Latin name Praesepe, the Manger. M44 covers 1.5 degrees of sky and contains some 50 stars of magnitude 6.0 and fainter. It lies approximately 590 light-years from Earth, and is best studied with binoculars.

M67 This is a larger but more distant cluster than the Beehive, lying roughly 2,300 light-years away. It contains about 200 stars within a half-degree of sky, and it shines with an overall brightness of magnitude 7.4, so it is not visible to the naked eye. Binoculars will show it as a dense elliptical patch of light, but a small telescope is needed to resolve individual stars.

132 Globular clusters

Far beyond the Milky Way, in a halo above and below our galaxy, orbit the globular clusters: huge intergalactic "glitter balls" that may contain millions of stars. More than 150 are known, and these strange outcasts are still revealing their mysteries.

A typical globular cluster contains several hundred thousand stars, crammed into a sphere just a hundred light-years in diameter, so their average separation is just a few light-days. Globular cluster stars are very different from those in our part of the galaxy. They are generally old, yellow or red, and made almost entirely of hydrogen and helium. It seems that most formed 10 billion or more years ago – before later generations of stars had a chance to turn much of the primordial gas into heavier elements. The lack of heavy elements prevents them following the most energetic hydrogen-burning reactions, and forces them into a slow, sedate life cycle.

Astronomers once thought that globular clusters were created at the same time as the galaxies; either in collisions between smaller clusters of young stars, or from small collapsing gas clouds. However, new globular clusters have recently been discovered around colliding galaxies, apparently in the process of forming.

Large black holes may have acted as the seeds for the creation of at least some globular clusters. A few certainly have active black holes at their hearts today, gobbling up matter and emitting fierce radiation. Some globular clusters also contain what appear to be young, blue, main-sequence stars, often concentrated near the centre. These blue stragglers seem to trail behind the rest of the cluster's evolution. They are probably the result of two elderly stars colliding in the crowded heart of the cluster to form a more massive, brighter and bluer star.

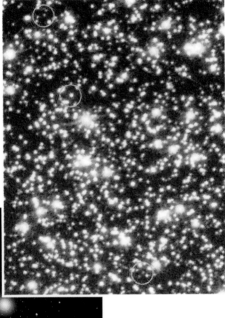

Going deeper into a globular cluster
As a globular cluster (left) grows older, more of its stars leave the main sequence and turn into red giants. Blue stragglers (above) tend to occur mostly in the crowded cores of globular clusters. They are probably ageing rapidly, dying and being replaced by new ones in a continuous process.

See also: Star birth 94 · Red & brown dwarfs 102 · Black holes 126 · Open clusters 130 · First stars & galaxies 176

Seeing clusters in Hercules

M92 This is the second, bright globular cluster in Hercules, slightly smaller and 2,000 light-years farther away than M13. It shines at magnitude 6.5, and is easily seen through binoculars. It is more tightly packed than M13, so fewer individual stars are resolvable except at its edges.

19 HERCULES *The sky's fifth-largest constellation is named after the ancient Greek hero who became immortal through performing 12 superhuman tasks. It is best seen around July and August. However, it is not as imposing as its size might suggest: the brightest stars are of magnitude 2.8 and fainter, and the pattern is not distinctive enough to locate easily in the sky. The centre of the constellation is made up of a quadrangle known as the Keystone. Hercules' arms and legs extend from the Keystone's four corners. From the northern hemisphere, he is seen upside down, so one knee rests on the head of Draco, the Dragon. Lying a little to the northeast of the Milky Way, the constellation contains several of the globular clusters that orbit above the galaxy's plane, including the very bright M13, many double stars and other interesting objects.*

M13 This is the finest globular cluster visible in northern skies. It contains about 300,000 stars in a sphere 110 light-years across, and lies 25,000 light-years from Earth. Shining at an overall magnitude 5.8, M13 is half the size of the Full Moon, and can be seen with the naked eye under the darkest skies. Through binoculars or a telescope at low magnification, it is a beautiful sight, with a loose structure that shows individual stars across most of its disc.

DQ Her The faint, 15th-magnitude star DQ Herculis, now well beyond the reach of small telescopes, underwent a nova outburst in 1934, shining at magnitude 1.3 for several weeks. It is a polar star – a magnetic white dwarf and a red dwarf star orbiting close to each other.

α Her Alpha Herculis is a double-star system containing one of the largest red giants known. Varying in brightness between magnitude 2.8 and 4.0, Rasalgethi – the Kneeler's Head – lies 380 light-years from Earth, and is orbited by a smaller red-giant star of magnitude 5.3, best seen in small telescopes.

STAR STORIES

The Epic of Gilgamesh

Today, Hercules is named after the legendary Greek hero and demigod. The constellation is supposed to represent the hero kneeling with his club raised ready to strike the head of Draco, the Dragon. However, this pattern of stars was identified with a kneeling super-man before Hercules' name was attached to it. To the peoples of ancient Mesopotamia, it represented the great hero Gilgamesh, who fought with the goddess Ishtar and sought to live forever.

The Milky Way

The faint band of light that runs across the night sky is made of countless distant suns. When we look into the Milky Way, we are looking across the plane of our own galaxy, a vast spiral of two hundred billion stars "floating" in intergalactic space.

The shape of the Milky Way tells us a surprising amount about our galaxy. It appears as a band across the sky because the Sun lies in the plane where most stars are concentrated. The brightest areas lie in Sagittarius, where vast numbers of stars hide our view of the galaxy's centre 26,000 light-years away. On the opposite side of the sky, we look out through a thinner region of stars, a spiral arm wrapping around the outside of the galaxy.

Star clusters arranged in chains mark out sections of the galaxy's spiral arms. Each is named after the region where it appears brightest – our Solar System lies in the Orion Arm.

Most of the stars in the spiral arms are relatively young and rich in heavy elements. Toward the centre, however, things are very different. The galaxy bulges out into a huge oval of very old, red and yellow stars – the nucleus. The spectra of these stars reveal that they are made almost entirely of hydrogen and helium and lack the heavier elements found in younger stars. This means they must have formed very

early in the galaxy's history, before such elements had been manufactured.

The Milky Way rotates (otherwise it would collapse under its own gravity) every 250 million years. But stars on the outer edges do not behave as if all the galaxy's mass lies within their orbits. This indicates that the visible Milky Way may be surrounded by a halo of invisible dark matter.

Rotation should also cause the spiral arms to wrap around the hub and disappear in a few turns. The arms can only survive because they are not true structures. Instead they are cosmic traffic jams where stars and nebulae are pushed together by the uneven pull of the galaxy's nucleus. This creates the bright knots of stars and bursts of star formation that define the spiral arms.

Perseus Arm This wraps around the side of the Orion Arm and contains constellations such as Perseus and Auriga.

Sagittarius Arm This blocks our view of the centre of the Milky Way, and contains huge amounts of stars in constellations such as Carina and Cygnus.

Outer arm

Spiral arms Each of the arm's leading edge is marked by regions of intense new star formation. Calmer regions containing more mature suns such as our own trail behind.

Orion Arm The Solar System lies on the Orion Arm, which contains many of the sky's brightest nebulae.

Our Solar System

Central bulge Evidence is mounting that the Milky Way is a barred-spiral galaxy with an elongated hub.

See also: · Our star, the Sun 36 · Stellar evolution 92 · The centre of our galaxy 136 · First stars & galaxies 176

The Milky Way and Puppis

75 PUPPIS *Representing a ship's stern, Puppis lies in a rich area of the Milky Way where the lack of nearby dust clouds allows us to peer into the depths of our galaxy. Here the outer edge of the Sagittarius Arm curves away from us, and in the foreground we see stars and clusters within our own local Orion Arm.*

STAR STORIES

Legends of the Milky Way

The Milky Way, the path of our galaxy, is undoubtedly among the most beautiful features of the night sky. Modern city lights hide this faint light, but over the centuries many wonderful stories and legends have been attached to it. For example, several ancient cultures have regarded the strip as a celestial river, often associated with a river on Earth, such as the Nile in Egypt, the Ganges in India and the Yellow River in China.

At one time, the people of Misminay in Peru believed that the Milky Way was some kind of heavenly channel, carrying water from the seas and rivers on Earth up into the sky, from where it would return to the planet as rain.

In other parts of the world, some cultures have been known to interpret the Milky Way as a stream of milk from the breast of a goddess; others have seen it as the path taken to the afterlife by dead souls.

M46 Fainter than M47, M46 is a cluster at the edge of naked-eye visibility. It contains more stars, but lies 4,600 light-years away.

M47 Although the star clusters M46 and M47 lie in the same part of the sky, they are in quite different parts of the Milky Way. M47 is a small but bright cluster, easily visible with the naked eye, and 1,600 light-years from Earth. It shines at magnitude 4.3 and contains about 30 stars. Both M46 and M47 are open clusters.

ξ Pup Xi Puppis is a yellow supergiant of magnitude 3.3, and lies 1,350 light-years from Earth.

M93 Puppis' third Messier object, M93 is a cluster of about 80 stars 3,600 light-years away. At magnitude 6.5, binoculars are needed to reveal its distinctive wedge shape.

ζ Pup One of the hottest stars known, Zeta Puppis is a brilliant, blue supergiant with a surface temperature of 35,000 °C (63,000 °F). It shines at magnitude 2.2 and is more than 1,400 light-years away.

NGC 2477 Another bright star cluster in Puppis is NGC 2477, a tight ball of stars shining at magnitude 5.7. Although it resembles a globular cluster, NGC 2477 is only 4,200 light-years away. It is therefore just a tightly bound, open cluster.

L¹,² Pup Through a small telescope, L Puppis appears to be a double star consisting of L¹, a blue-white, magnitude-4.9 star about 150 light-years away, and its red companion L². The alignment is pure coincidence, however, and L² is actually a red giant 190 light-years from Earth. It is also a pulsating variable star, cycling between magnitude 2.6 and 6.2 in approximately 140 days.

136 The centre of our galaxy

Deep in Sagittarius, 26,000 light-years away, lies the most violent place in our cosmic neighbourhood – home to enormous stars, fountains of antimatter and a giant black hole weighing more than two million Suns.

In the centre of the Milky Way, stars are closely packed together, and may even collide at times, merging to form stellar giants that break all the rules of normal stellar evolution. Most of the stars in the central nucleus are old and red. They formed early in the galaxy's history and have few of the heavy elements that help to speed up the life cycles of younger stars. These stellar fossils are next door to much younger neighbours – the gas and dust clouds in the nucleus are production lines for clusters of bright, massive stars.

The galaxy's core is hidden by a ring of dense clouds surrounding a central cavity ten light-years across that completely block visible light. By using infrared light, however, astronomers can see stars at the very heart of the Milky Way. Their movements show that they are being affected by an object with 2.6 million times the mass of the Sun, crammed into a space just one-thirtieth of a light-year across. With such incredible density, this object can only be a giant black hole.

Today, this black hole is asleep. The area within its gravitational reach has been swept clear of material, and the giant stars around it keep their distance. A little dust still falls steadily inward, heating up and emitting radio waves. But things were not always so quiet. Thousands of light-years above the nucleus lies a huge fountain of antimatter, glowing with high-energy gamma rays. The antimatter was probably blasted out of the nucleus the last time that an ill-fated star drifted into the black hole's grasp, and the sleeping giant awoke.

Fire in the darkness This image shows how the central Milky Way would appear if our eyesight was shifted into the infrared, allowing us to peer through the dark dust clouds that surround the nucleus. Every single speck of light is an individual star, and towering filaments of gas rise high above the horizontal plane of the Milky Way, following the lines of the galaxy's magnetic field.

See also: Star distances 84 · Star colours 88 · Stellar evolution 92 · Red & brown dwarfs 102 · Stellar heavyweights 104

Viewing the heart of the Milky Way

78 SAGITTARIUS *This group of stars, representing an archer, is a zodiac constellation best viewed when the Sun is on the opposite side of the sky: in summer from the northern hemisphere, and in winter from the southern hemisphere. Although the constellation stretches across a large area of sky, the brightest stars are clustered together in its northwest corner, with the constellation's most interesting objects around them. A narrow but bright strip of the Milky Way also runs through this corner. Here you are looking straight across the galaxy's spiral arms where the density of stars is at its highest. Beyond this bright strip lies the darkness of the Great Rift, an enormous dust cloud that obscures the galaxy's exact centre, just to the south of Beta Sagittarii, the constellation's most easterly bright star.*

STAR STORIES

The Celestial Archer

The constellation Sagittarius was identified as an archer on horseback (the god Nergal) by Mesopotamian astronomers as early as the 11th century B.C. By ancient Greek times, the figure had transformed into a centaur, half man, half horse, and this association has stuck, even though there is another centaur visible in southern skies. According to the Greeks, Sagittarius represented the wise centaur Chiron. He is about to avenge the death of the hunter Orion by firing an arrow at the heart of Scorpius.

M17 With binoculars, M17, the Omega Nebula, appears as a faint streak of light. Medium-sized telescopes are needed to show that it actually has a horseshoe shape, like the Greek letter Omega.

M23 M23 is a very bright, open cluster of young stars lying 2,100 light-years away. Shining at magnitude 5.9, it is best seen with a telescope.

M22 M22 is a bright globular cluster, one-third of a degree across and visible to the naked eye as a fuzzy star at magnitude 5.1. It is actually a huge ball of stars 10,400 light-years away, but a telescope is needed to separate individual stars from the mass.

M20 Messier 20, the Trifid Nebula, lies 1.5 degrees north of M8, and is actually in the same region of space. M20 appears as a blur in a small telescope, but long-exposure photographs show its details beautifully.

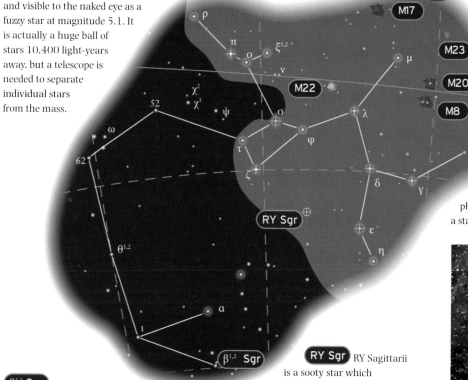

M8 The Lagoon Nebula, M8, is one of the brightest nebulae, visible to the naked eye, but best seen through binoculars or a small telescope. It covers an area three times the size of the Full Moon and looks milky-white, with a dark rift down the centre. Long-exposure photographs pick up the nebula's pinkish colour. M8 is a star-forming region more than 5,000 light-years away.

RY Sgr RY Sagittarii is a sooty star which occasionally plunges in brightness from magnitude 6.8 (just below naked-eye visibility) to magnitude 14.0, beyond most small telescopes.

β¹·² Sgr Beta Sagittarii consists of two unrelated stars of around magnitude 4.0. Arkab Prior is a blue-white, main sequence star 380 light-years away, whereas Arkab Posterior is a white star 139 light-years from Earth.

The great beyond Two spiral galaxies collide at speeds of hundreds of kilometres per second in one of the universe's most violent events. Stars in the delicate spiral arms are flung away into the dark intergalactic void, while gas clouds collide together in a burst of star formation. Over many millions of years, the central nuclei of these galaxies may fall together and eventually merge, but collisions between individual stars will be rare.

Beyond the Milky Way

140 Satellite galaxies

Our galaxy does not travel alone through the cosmos but holds several small irregular galaxies in orbit around it. Although not as large as the Milky Way, these smaller galaxies are far larger than the biggest globular or open star clusters.

By far the best known of our galaxy's satellites are the Magellanic Clouds – two irregularly shaped clouds of gas, dust and stars visible in southern skies as detached wisps of the Milky Way. Astronomers realized that they were separate galaxies in their own right only in the early 20th century, and until quite recently these were the only galaxies known to orbit our own.

The Large and Small Magellanic Clouds orbit our galaxy once every 1.5 billion years, travelling on an elliptical path that brings them to within about 120,000 light-years of the Milky Way at their closest approach, and takes them out to 400,000 light-years at their most distant. With each close approach, our galaxy's enormous gravity tugs at them, pulling away material that trails round the orbit behind them. Hydrogen in this Magellanic Stream emits radio waves that mark the path of the clouds around the sky.

The Large Magellanic Cloud (LMC) shines as brightly as 2 billion Suns, and currently lies about 150,000 light-years from Earth in the constellation of Dorado. Its distance has only recently been accurately fixed, based on calculations of the real brightness of eclipsing binary stars within it. The LMC is currently on the outward leg of its orbit, having passed closest to the Milky Way a quarter of a billion years ago. It is about one-twentieth the mass of our galaxy, and has very little structure. A bright bar of stars 10,000 light-years long runs across its centre, with a vague spiral arm emerging from one end. It is peppered with bright star clusters and nebulae, including the enormous Tarantula Nebula.

The Small Magellanic Cloud (SMC) lies 30,000 light-years farther from Earth, and contains only a quarter as many stars as the LMC. It has no internal structure, but a distinctive peanut shape caused by the pull of the Milky Way. An enormous burst of star formation must have occurred here during its closest approach to our galaxy around 100 million years ago, as more than half the SMC's bright star clusters date to this time.

Until 1994, the Magellanic Clouds were thought to be the closest galaxies to our own. But then, behind the star clouds at the centre of the galaxy, a small galaxy was noticed hiding above the plane of the Milky Way, just 80,000 light-years from Earth. The Sagittarius Dwarf Elliptical Galaxy (SagDEG for short) has a spherical structure like a giant globular cluster, distorted into a teardrop shape as our galaxy's gravity gradually strips away its stars.

Behind the Milky Way's dusty gas lurk several other faint dwarf ellipticals. It now seems that our galaxy is surrounded by a small swarm of satellites, but most of these are too faint to see except with powerful telescopes or at wavelengths other than visible light.

Magellanic Clouds The Milky Way with its major satellite companions, the Large and Small Magellanic Clouds.

See also: The starry sky 82 · The Milky Way 134 · The Local Group 144 · Distances to galaxies 146 · Types of galaxy 148

Goldfish galaxies

54 **DORADO** *The small faint constellation of Dorado, the Goldfish – sometimes also known as the Swordfish – lies west of Carina, with its bright star Canopus, and the Milky Way. Although the constellation itself is relatively uninteresting, it does contain one of the finest sights in southern skies: the Large Magellanic Cloud. Dorado rises highest around October (spring in the southern hemisphere) but lies so far south that it can only be seen from equatorial regions in the northern hemisphere.*

β Dor Beta Doradus is much closer than the LMC, just 1,000 light-years away. It is a yellow supergiant star, and a bright example of a Cepheid variable star. Its brightness ranges from magnitude 3.5 to 4.1 as it pulsates in a 9.8-day cycle.

NGC 2070 The Tarantula Nebula is a vast emission nebula, 800 light-years across – far larger than any star-forming region in the Milky Way. The nebula is lit from within by R136, a cluster of young supergiant stars. To the naked eye, it appears as a fuzzy star, but binoculars transform it, revealing the tendrils of gas that produce its spiderlike shape.

α Dor The brightest star in Dorado, Alpha Doradus, is a blue-white giant star of magnitude 3.3, 175 light-years from Earth.

SN 1987A In 1987, a supernova explosion lit up the LMC on the edge of the Tarantula Nebula. This was the brightest supernova seen from Earth in nearly 400 years, reaching magnitude 2.8. It remained visible to the naked eye for ten months.

NGC 1566 Far more distant than the Magellanic Clouds, Dorado contains several other galaxies. The brightest is NGC 1566, easily visible with a small telescope. This is a face-on spiral Seyfert galaxy, with an unusually bright nucleus.

LMC The Large Magellanic Cloud (LMC) is easily visible with the naked eye, looking like an isolated chunk of the Milky Way about 10 degrees across and 20 times wider than the Full Moon. Its diameter is about 30,000 light-years. Binoculars reveal fuzzy stars scattered across its face, each of them in reality an entire open star cluster.

✦ STAR STORIES

Magellanic Legends

The astronomers of other cultures have often explained the Magellanic Clouds with names and stories. The Karanga Tribe of southern Africa called them "Plenty" and "Famine", and the Australian Aborigines believed that the clouds of stars had been torn away from the Milky Way.

142 The Andromeda Galaxy

Beyond our Milky Way, a vast gulf of space some 2 million light-years wide separates us from our nearest large galactic neighbour, the Andromeda Galaxy. This huge stellar system is a spiral galaxy like the Milky Way, but even larger.

The Andromeda Galaxy, often referred to by its Messier number, M31, lies 2.5 million light-years from the Milky Way. This is surprisingly close, considering the size of the two great galaxies—the gap is barely 20 times the diameter of the Milky Way.

Despite its proximity, large telescopes are needed to resolve the galaxy's individual stars. Astronomers have measured the distance to M31 by looking for certain standard candles – objects that have the same brightness wherever they occur. These include Cepheid variable stars, some types of supernovae and even star-forming hydrogen clouds. The galaxy they reveal is very like our own. Toward its centre, M31 bulges outward and brightens as the number of stars increases. The reddish-yellow light from this region reveals that M31, like the Milky Way, contains two distinct types of stars: ancient, slow-burning stars at the hub; and young, fast-evolving stars, rich in heavy elements, in the spiral arms.

An estimated 400 billion stars are contained in M31, twice as many as our galaxy. At 150,000 light-years across, Andromeda is also 50 per cent wider than the Milky Way. The Doppler shift of M31's light indicates that it is moving toward us, and will collide and merge with our own galaxy in billions of years' time. One edge also appears to be moving toward us faster than the other, indicating that M31 is rotating.

A swarm of satellite galaxies surround M31, most notably a pair of dwarf ellipticals hanging above its central plane. But the objects under M31's influence behave as if they are orbiting a galaxy that is lighter than the Milky Way, despite the fact that M31 contains more stars. This is probably because both M31 and the Milky Way contain far more mass than is accounted for by their visible stars. This nonvisible mass, or dark matter, gives our galaxy an overall mass of 2 trillion Suns, whereas Andromeda has only 1.2 trillion Suns' worth of material. In both galaxies, the dark matter extends for hundreds of thousands of light-years beyond the outermost visible stars and is responsible for keeping the outer reaches of each galaxy spinning rapidly.

A galaxy like our own The Andromeda Galaxy lies almost edge-on to us, and we see it from just above its central plane. The spiral arms are hard to distinguish, but dark dust lanes around their edges - huge chains of dark nebulae - help to define the structure.

See also: The Milky Way 134 · Satellite galaxies 140 · The Local Group 144 · Types of galaxy 148

Andromeda –
the maiden in chains

1 ANDROMEDA *Lying well away from the Milky Way, the relatively faint constellation of Andromeda offers us a clear view of deep intergalactic space, without obscuring dust and stars. Andromeda lies to the south of Cassiopeia, shares its brightest star with the Square of Pegasus, and contains just a few interesting objects. However, one of these is M31, the Andromeda Galaxy, and the most distant object visible with the naked eye. Andromeda was the daughter of Cassiopeia and Cepheus and she was chained to a rock in the sea as a sacrifice to a sea monster, but was rescued by the hero Perseus.*

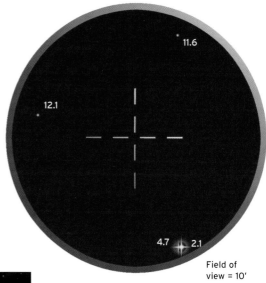

Field of view = 10'

M31 From a dark site, the Andromeda Galaxy, M31, is easily visible with the unaided eye, shining at magnitude 4.3. Binoculars reveal the galaxy as an oval smudge with a bright centre and fainter nebulosity around the edges. A small telescope will show more detail, and may allow you to see Andromeda's two small companion galaxies – M32 and NGC 205.

γ And Gamma Andromedae is a multicoloured triple star, 355 light-years from Earth. Binoculars or a small telescope will easily show yellow and blue stars of magnitudes 2.1 and 4.7. Higher magnifications can reveal that the yellow star also has a companion.

NGC 7662 This planetary nebula, known as the Blue Snowball, is relatively bright at magnitude 9.2, but too compact to reveal details except at high magnification. Through binoculars, it appears as a fuzzy blue-green star, but a telescope shows a hazy, elliptical ring. In fact, NGC 7662 has two rings around it: a bright inner one and a much fainter outer one.

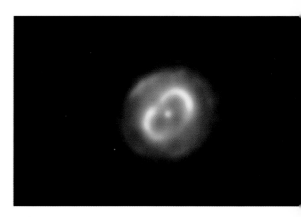

NGC 752 This open cluster of about 60 stars is 1,300 light-years from Earth. It shines at an overall magnitude 6.6 and is easily visible with binoculars, covering an area 50 per cent wider than the Full Moon.

α And Alpha Andromedae, or Alpheratz, on Andromeda's border with Pegasus, is a hot blue-white star of magnitude 2.0, 97 light-years from Earth.

STAR STORIES

A Nebulous Galaxy

The Islamic observer Al-Sufi first recorded the Andromeda Galaxy in A.D. 964, and the German astronomer Simon Marius studied it with one of the first telescopes in 1612. Astronomers speculated that this fuzzy patch of light might be a hole in the heavens or a glowing patch of ether. In 1845, William Parsons (1800-1867), the Earl of Rosse, using his 90-cm (36-inch) telescope, saw star clusters inside it and recognized that starry nebulae were very different from gas clouds such as the Orion Nebula.

144 The Local Group

The Milky Way is one of three large spiral galaxies in an area about three million light-years across. With other irregular and dwarf galaxies, these three make up the Local Group – a small cluster of galaxies that is our immediate cosmic neighbourhood.

The Local Group contains some 32 galaxies. Of these, 11 are irregular clumps of gas and young stars like the Magellanic Clouds, and 18 are dwarf ellipticals, rather like oversized but sparse globular clusters. The other three, which contain over 95 per cent of the visible material in the Local Group, are the Milky Way Galaxy, the Andromeda Galaxy – M31 – and the Triangulum Galaxy – M33. Most of the galaxies in the Local Group are clustered together in two regions surrounding the Milky Way and the Andromeda Galaxy.

The Triangulum Galaxy is much smaller than either of these great spirals, but it, too, has a spiral structure, displayed clearly because it happens to lie face on to Earth. Although looser and fainter than Andromeda, M33 has the same types of stars, with old red and yellow stars at its hub, and young, blue stars in its spiral arms. It also contains one of the largest star-forming regions known: the gigantic emission nebula NGC 604, 15,000 light-years across.

Almost half of the smaller galaxies in the Local Group are satellites of either the Milky Way or the Andromeda Galaxy. M33 seems to be too small to have its own satellites.

The Magellanic Clouds are by far the biggest of the Local Group's minor members. The other irregulars are smaller versions of these clouds – rich in gas, young stars and star-forming regions. Most of the dwarf ellipticals are loose collections of old stars, with the space in between the stars so empty of interstellar gas that it is possible to look right through them.

We may never know the true number of galaxies in the Local Group. This is partly because the dwarf ellipticals are so faint that they are almost impossible to detect beyond the Andromeda Galaxy, but mostly because the star-clouds of our own galaxy form an obscuring band around the sky, blocking off much of our view. This is why the closest galaxy to our own, the Sagittarius Dwarf Elliptical, barely 80,000 light-years away, was not discovered until 1994.

As in individual galaxies, the visible material in the Local Group is only half the story. By studying the movements of our neighbouring galaxies, astronomers have estimated that the Local Group is up to 20 times more massive than all the stars, gas, dust and other visible material within it would suggest. The rest must exist within and around the galaxies as mysterious dark matter.

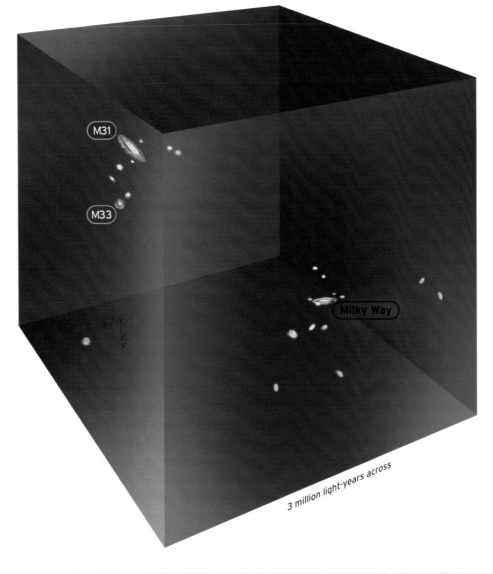

3 million light-years across

Neighbourhood map Apparently empty space surrounds our immediate corner of the Universe until the nearest groups of galaxies, around 10 million light-years away.

See also: The Milky Way 134 · The Andromeda Galaxy 142 · Types of galaxy 148 · Elliptical galaxies 150

The galactic neighbourhood

33 **3** **TRIANGULUM & ARIES** *Triangulum and its neighbouring constellation Aries, the Ram, lie in a relatively barren region of sky between Taurus and Andromeda. The Milky Way runs north of both constellations, so the skies reveal clear vistas of intergalactic space. Aries is a zodiac constellation where the planets of the Solar System often appear. Once, the Sun crossed the celestial equator from south to north here, marking the Vernal Equinox and the start of northern spring. Precession has now carried this point, the zero mark of the celestial coordinate system, into the neighbouring constellation of Pisces.*

α Ari Alpha Arietis, or Hamal, is the brightest star in this region of the sky. Shining at magnitude 2.0, it is a yellow giant, 66 light-years from Earth.

γ Ari Gamma Arietis, or Mesarthim, is an attractive double star consisting of twin, magnitude 4.8, blue-white stars, easily separated with good binoculars or a telescope. The stars lie 200 light-years from Earth, and each shines as brightly as 25 Suns.

M33 The Triangulum Galaxy is the closest face-on galaxy to Earth, roughly 2.5 million light-years away. Its closeness to the neighbouring Andromeda spiral is, for once, real, and not a line-of-sight effect. However, the galaxy is a relatively disappointing sight through small telescopes. Although it has an overall brightness of magnitude 6.2, its large size means it is spread out and has little contrast with the background sky. M33 is also naturally fainter than the Andromeda Galaxy, and lacks a bright central nucleus. So with binoculars it can be a challenge to see it at all.

λ Ari Double star Lambda Arietis consists of a white magnitude 4.8 star and its magnitude 7.3 yellow companion. They are 133 light-years away, and are visible through binoculars.

NGC 772 This is the brightest and most easily seen of many galaxies in Aries. It shines at magnitude 11.1 and lies well outside the Local Group, but is relatively easy to find with a small telescope because of its bright nucleus. Small instruments show it as a faint oval smudge of light with a brighter spot of light in the centre. Larger telescopes are needed to show its spiral structure.

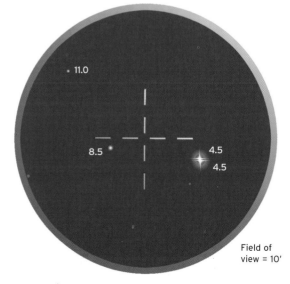

Field of view = 10′

146 Distances to galaxies

Beyond the Local Group, galaxies stretch away to the edges of the Universe. Even the closest galaxies are far too distant to measure using direct parallax methods, but astronomers still want to find out as much as they can about them.

Until the 1920s, most astronomers believed that our own galaxy was the limit of the Universe. The various spiral and other nebulae observed in different regions of the sky were believed to lie in the halo around our galaxy, or to be, at most, star clusters in distant orbits around the Milky Way.

The US astronomer Edwin Hubble was responsible for destroying this view of the cosmos. He searched the nebulae for standard candles, the celestial objects with a predictable luminosity. The magnitude of such an object as seen from Earth shows how far away it is. Hubble used Cepheid variables – pulsating stars whose period of pulsation is related to their real brightness. By identifying Cepheids in other galaxies from their distinctive brightness changes, Hubble was able to find their true luminosities, and therefore distances.

But Cepheids are too faint to be seen in very distant galaxies, beyond about 50 million light-years from Earth. At greater distances, astronomers have to use other standard candles. It seems that the brightest globular clusters around a galaxy never exceed a certain luminosity, so astronomers can look for the brightest of these clusters and assume that they have that fixed luminosity. A similar method has been used with giant emission nebulae, such as the Large Magellanic Cloud's Tarantula Nebula, which also seem to have an upper limit to their size and brightness.

When even globular clusters become invisible, there is another relationship that can stretch the distance chain farther. This is a surprisingly simple link between the rotation rate of a spiral galaxy and its luminosity. By measuring the Doppler shifts of light from a galaxy as it rotates, it is easy to calculate the speed of rotation, its true brightness and therefore its distance.

Splitting the light from galaxies of known distance and studying their spectra reveals something remarkable: the light from all distant galaxies is shifted toward the red, implying that they are moving away from us. Even more surprisingly, the farther away a galaxy is, the faster it seems to be moving away (Hubble's law). In fact, not everything is moving away from us, but our galaxy is part of a general expansion, and everything is moving away from a point of origin somewhere in the distant past.

An expanding Universe can explain many mysteries, such as why all the galaxies do not collapse together under their own gravity. Most astronomers now accept that redshift means the cosmos is expanding, and this discovery means that redshift itself can be used as a measure of distance. By this method, astronomers have found the distances to galaxies billions of light-years away. Because the light from these objects takes so long to reach Earth, we are often seeing them as they were aeons ago, when the Universe was young.

Nearby galaxy The galaxy M82 lies 10 million light-years away in Ursa Major. Astronomers can estimate its distance by the brightness of the largest star-forming clouds within it, and the globular clusters that surround it.

See also: Star distances 84 · Pulsating stars 114 · Globular clusters 132 · The expanding cosmos 166

Clusters and clouds in Tucana

47 Tuc One of the sky's best globular clusters is 47 Tucanae. Easily visible to the naked eye as a magnitude 4.0 star, it is 15,000 light-years away. Binoculars reveal it as a densely packed ball of light, and small telescopes will resolve individual stars around its edges.

β^{1,2,3} Tuc Through binoculars, the multiple star Beta Tucanae appears as a pair of magnitude 4.2 and 4.4 blue-white stars. But a nearby magnitude 5.0 white star is also a member of the system, and one of the brighter components is a close double.

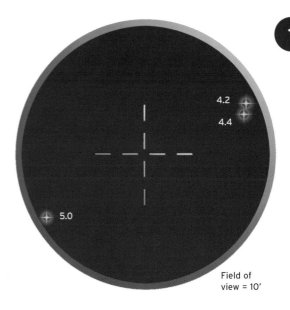

Field of view = 10'

85 **TUCANA** *Lying close to the South Celestial Pole, the constellation Tucana, the Toucan, is a faint and undistinguished group of stars. But it contains two objects of interest to any observer: the bright globular cluster 47 Tucanae and the Small Magellanic Cloud (SMC), one of the galaxies closest to our Milky Way.*

α Tuc The constellation's brightest star, Alpha Tucanae, is an orange giant of magnitude 2.9. It is 200 light-years from Earth.

SMC The Small Magellanic Cloud is an irregular galaxy an estimated 180,000 light-years away. To the naked eye, it looks like a detached, peanut-shaped cloud of the Milky Way. Its distance from Earth has been calculated based on Cepheid variables. In fact, the Cepheid period-luminosity relationship was first discovered in the SMC.

NGC 362 Globular cluster NGC 362 is smaller and fainter than 47 Tucanae. It is also twice as distant. Nevertheless, shining at magnitude 6.5, it is worth a look through binoculars or a small telescope.

NGC 346 Visible through binoculars as a brighter patch of the Small Magellanic Cloud, NGC 346 is one of the largest known star-forming regions. At its heart lies a magnitude 10.3 star cluster, best seen with a small telescope.

STAR STORIES

Magellan

The first European to record the Magellanic Clouds was the Portuguese navigator Ferdinand Magellan (c. 1480-1521). He led the first expedition to circumnavigate the globe from 1518-1522, passing round the Cape of Good Hope and through the Magellan Strait. Although Magellan did not survive the voyage, his discoveries made a lasting contribution to science.

148 Types of galaxy

The Universe contains at least as many galaxies as there are stars in the Milky Way, but there are only a few basic types. The classification system for galaxies was invented by Edwin Hubble, who first showed that galaxies lay beyond the Milky Way.

Hubble's classification, often called the tuning fork diagram, divides galaxies into four main types: spirals, barred spirals, ellipticals and irregulars. Spirals and barred spirals are galaxies similar to our own. They have a central bulge of old red stars, surrounded by a disc of younger ones. A barred spiral's bulge is distorted from a sphere into a long ovoid, or bar, with the spiral arms at either end. The bar may form when stars are thrown into eccentric elliptical orbits (the Milky Way is now thought to be a barred spiral, with its bar pointing straight toward Earth). Spiral arms of bright blue stars running across the disc mark slowly rotating regions where waves of compression actively form stars. The lives of the galaxy's brightest, most massive stars are so brief that they accurately mark the compression region's current position.

Most spirals are smaller than the Milky Way and Andromeda Galaxy and many have much looser structures – the regions of star formation are not so confined, and the spiral is broken up. These flocculent spirals are believed to be systems in which the spiral wave of compression and star formation is relatively weak.

Irregular galaxies share many of the features of spirals, with large amounts of gas and dust, young stars and huge star-forming regions. Many show signs of central bars and even the beginnings of spiral arms. They seem to be galaxies too small to form into proper spirals.

Elliptical galaxies are simply huge collections of old red stars, ranging from faint dwarfs far smaller than our galaxy, through normal types with as many stars as a spiral, to giant versions found at the centre of huge galaxy clusters.

Lenticular galaxies, halfway between ellipticals and spirals, seem to be spirals without the spiral arms. They have a central hub of old stars and a sparse disc with a few younger, Sunlike stars, but none of the star-forming regions and massive stars of spiral galaxies. They appear to be spirals that have somehow lost the interstellar gas that would allow them to form new stars.

Normal spirals

Sc

Sb

Sa

The letter a, b, or c describing a spiral or barred-spiral galaxy indicates the size of the nucleus and spiral arms. Sa and SBa galaxies have the largest nuclei and the tightest spiral arms.

Barred spirals

SBc

SBb

SBa

SO or lenticular - a galaxy with a spiral-like hub and disk, but no arms.

Ellipticals

E7 ellipticals are the most elongated.

E4 ellipticals have noticeably stretched shapes.

E0 galaxies are perfectly spherical balls of stars.

The number given to an elliptical galaxy indicates just how elliptical it really is. The exact shape we see depends on the angle at which we view the galaxy, but E0s, at least, appear to be truly spherical.

Hubble's tuning fork diagram Each type of galaxy is given a letter and a number to indicate its exact shape. Hubble thought at first that one type of galaxy evolved into another. We now know that things are not this simple, but the diagram is still useful for showing the variety of different shapes.

See also: The Milky Way 134 · Distances to galaxies 146 · Elliptical galaxies 150 · First stars & galaxies 176

The galaxies of Fornax

NGC 1316 This strange, barred spiral of magnitude 9.3 is a nearby example of an active galaxy. It contains more dust than its neighbours, and has an unusually bright central nucleus – so it is classed as a Seyfert galaxy. A small telescope will reveal the galaxy's disc as a faint oval of light surrounding the nucleus. NGC 1316 is associated with a strong radio source called Fornax A.

α For The constellation's brightest star is magnitude 3.9 Alpha Fornacis, 46 light-years from Earth. A telescope reveals that the star is binary, with the yellow primary orbited by a yellow-orange companion of magnitude 6.9.

Fornax This constellation has a secret: it hides a galaxy cluster with 18 bright members and at least 10 fainter ones. The cluster is a real association of galaxies bound together by gravity, not a chance line-of-sight effect.

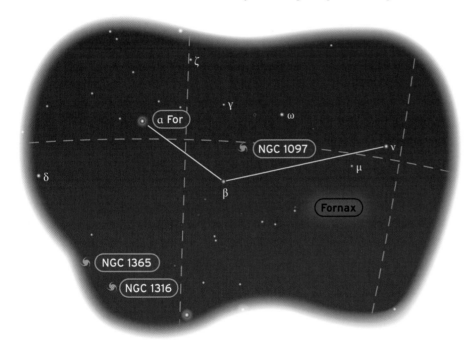

56 FORNAX *Lying to the south of Cetus, and largely encircled by the winding constellation of Eridanus, Fornax, the Furnace – first christened Fornax Chemica, the Chemist's Furnace – is an inconspicuous group of stars, none brighter than 4th magnitude. Fornax rises highest in the sky in the evenings around November, when a small telescope and a dark sky can reveal galaxies of several different types.*

NGC 1097 At an overall brightness of magnitude 10.3, NGC 1097 is one of the sky's best barred spirals (type SBb), face on to Earth, and roughly 60 million light-years away. Like NGC 1365 (left), a small telescope will clearly show the galaxy's bright nucleus in a surrounding oval halo of light. Good viewing conditions or a slightly larger instrument should reveal the spindlelike bar running through this halo.

NGC 1365 This attractive barred spiral (type SBb), of magnitude 10.3 is easily visible through small telescopes. However, its internal structure (left) is more difficult to detect and really needs a medium-sized instrument.

Elliptical galaxies

150

Elliptical galaxies include the smallest and the most massive of galaxies. These stellar groupings are made almost entirely of old, red stars, and show little sign of ongoing star formation. They probably account for most of the galaxies in the Universe.

Elliptical galaxies are collections of stars of various sizes, ranging from the dwarf ellipticals that are hard to see even when nearby, to giants whose gravity reigns over entire clusters of galaxies. More normal ellipticals are of similar mass to the Milky Way, but their stars orbit in an elongated ball rather than a disc. Although ellipticals only account for 10 per cent of bright, nearby galaxies, the huge number of faint dwarfs means that ellipticals as a whole may outnumber all other galaxies. All ellipticals have certain features in common, such as a lack of gas and hot, young stars.

Elliptical galaxies are classified according to their shape, and range from perfectly spherical to highly elongated. The orbits of their stars create the galaxies' shapes. Astronomers can work out trends in stellar orbits in these distant galaxies by analysing their spectra. Because some stars will be moving away from us, and some coming toward us, the dark absorption lines in a spectrum will be both redshifted and blueshifted by the Doppler effect.

It has been found that ellipticals in general rotate far more slowly than spiral galaxies, and that the movements of stars in spherical galaxies are totally random. The more elliptical a galaxy becomes, the more ordered patterns begin to appear in the orbits of its stars.

Dwarf ellipticals are small clusters of stars, and may be survivors from some of the first small

gas clouds in the Universe. The stars formed in these clouds would be almost pure hydrogen and helium. Their purity would help to extend their lifetimes, but the most massive stars would still consume all their fuel in a few million years, swelling to red giants before blowing apart in supernova explosions.

The gravity of a dwarf elliptical is so weak that the force of an exploding star would completely overcome it. Material from a supernova would simply blow away into intergalactic space, and the shockwave would disperse any other gas left in the cluster. Just a few supernovae could strip a dwarf elliptical of its gas, leaving it dying with just a few ancient, low-mass stars still burning away.

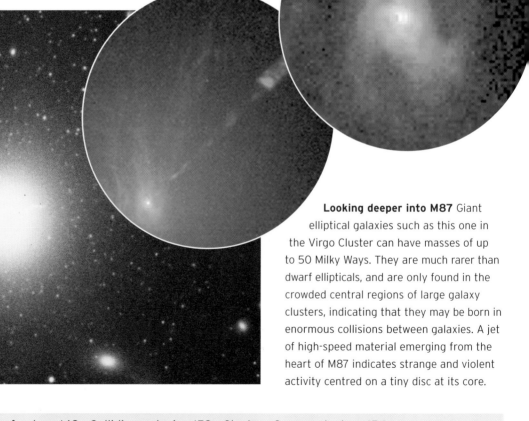

Looking deeper into M87 Giant elliptical galaxies such as this one in the Virgo Cluster can have masses of up to 50 Milky Ways. They are much rarer than dwarf ellipticals, and are only found in the crowded central regions of large galaxy clusters, indicating that they may be born in enormous collisions between galaxies. A jet of high-speed material emerging from the heart of M87 indicates strange and violent activity centred on a tiny disc at its core.

See also: Supernovae 120 · Types of galaxy 148 · Colliding galaxies 152 · Clusters & superclusters 154

The realm of elliptical galaxies

The zodiac constellation Virgo, the Virgin, represents either a goddess of justice, or the Greek harvest-goddess Demeter. It lies to the south of Boötes, on the celestial equator, and is easily seen from both hemispheres. It is also a good distance away from the Milky Way, affording a clear view through deep space to the nearest major cluster of galaxies. This grouping, the Virgo Cluster, contains about 3,000 galaxies and lies about 45 million light-years from Earth.

NGC 4261 The 11th-magnitude elliptical galaxy NGC 4261, 100 million light-years from Earth, hides a secret at its centre: a black hole surrounded by a disc of dusty material. As the black hole pulls material inward, it ejects jets of superheated gas and dust from above and below, as seen in this artist's impression of the view from an asteroid in the plane of the disc. The disc may be all that remains of a smaller galaxy that was swallowed by NGC 4261 hundreds of millions of years ago.

M104 The spiral galaxy M104, known as the Sombrero Hat Galaxy, is one of the most distinctive in the sky. From our nearly edge-on viewpoint, the galaxy, about 35 million light-years away and outside the Virgo Cluster, shines at magnitude 9.2. Its outer dust lanes are clearly visible through small- and medium-sized telescopes.

γ Vir The double star Gamma Virginis consists of twin yellow stars of magnitude 3.5, shining together at a total magnitude 2.8. The stars lie about 39 light-years away, and orbit each other every 169 years. For a few years around 2005, they will be too close to separate with small telescopes.

M87 Shining at magnitude 8.1, this giant elliptical galaxy in the Virgo Cluster is an almost perfectly circular ball of light when seen through small telescopes.

M49 Another elliptical member of the Virgo Cluster, Messier 49 shines at magnitude 9.3, at the limit of binocular visibility but it is easily seen in small telescopes, and appears considerably more elongated than M87.

α Vir Alpha Virginis, also known as Spica – the Ear of Wheat – is a blue-white giant star 260 light-years from Earth. It shines at magnitude 1.0 and is a binary whose components are too close to separate.

152 Colliding galaxies

Not all galaxies fit spiral, lenticular or elliptical classifications. Some have huge streamers of stars emerging from them or strange cartwheel structures. Such galaxies are in the process of colliding and evolving new forms before our eyes.

On the grand scale, galaxies are closely packed, typically separated by just a few times their own diameter. As a result, they exert an influence on each other, falling inward and sometimes colliding in the most spectacular cosmic events.

Galaxy collisions and near-misses are quite common, but even when two galaxies meet head-on, their individual stars rarely collide. However, the gas clouds of each galaxy slam into each other at speeds of 300 km per second (180 mps) or more. This compresses and heats them up, often triggering brief but enormous episodes of star formation, or starbursts. Because most galaxies seem to have large halos of dark matter extending well beyond their visible size, even an apparent near miss can create a starburst galaxy.

Collisions strip galaxies of their gas, leaving them unable to form any more stars after the initial starburst. The rise in temperature and pressure allows gas to reach speeds high enough to escape the galaxies, although still remaining trapped within a galaxy cluster.

The enormous gravitational and tidal forces involved can pluck stars from their orbits and fling them into intergalactic space, where they may form long chains. If two galaxies collide at right angles, the spiral structure can be destroyed, and a new wave of star formation can form a ring galaxy with dense star formation at the centre and in an expanding ring outside.

As the stars that survive coalesce in the aftermath, they may form into a larger elliptical galaxy. This seems to explain why ellipticals are stripped of their gas and no longer form new stars. Some ellipticals have been seen apparently absorbing other galaxies, and long-exposure photographs reveal that they are surrounded by faint rings of stars in a halo.

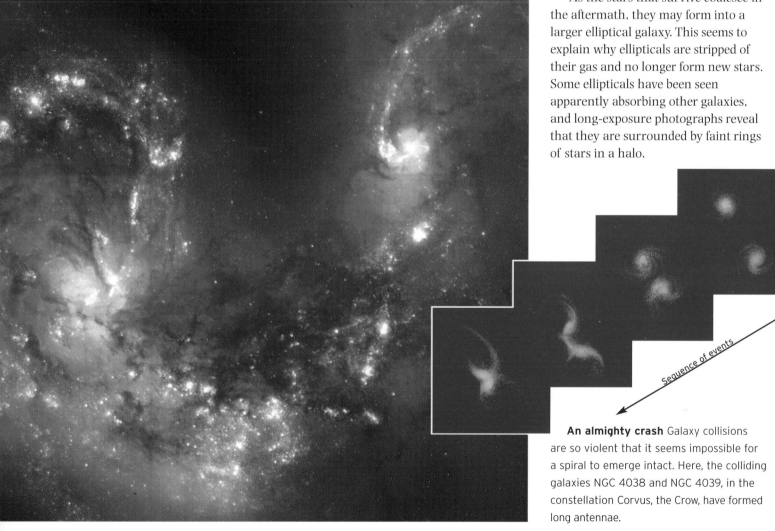

Sequence of events

An almighty crash Galaxy collisions are so violent that it seems impossible for a spiral to emerge intact. Here, the colliding galaxies NGC 4038 and NGC 4039, in the constellation Corvus, the Crow, have formed long antennae.

See also: Types of galaxy 148 · Elliptical galaxies 150 · Clusters & superclusters 154 · First stars & galaxies 176

Whirlpools in space

8 CANES VENATICI *The northern constellation of Canes Venatici represents the hunting dogs of the herdsman Boötes, eternally chasing the Great and Little Bears around the pole. The constellation lies to the south of Ursa Major, below the handle of the Plough. Its stars are relatively faint and sparse, but it contains a good collection of galaxies, most of which belong to the nearby supercluster centred in Virgo. It also contains the finest example of interacting galaxies visible to a small telescope.*

M51 With its open face displayed toward Earth, the Whirlpool Galaxy is probably the finest spiral galaxy in the sky, although it is best seen through a medium-sized telescope. It shines at magnitude 8.7, and lies 15 million light-years away. It is accompanied by NGC 5195, a small distorted spiral galaxy that lies just to the north (at the bottom of this telescopic image) and slightly farther away. The galaxies had a close encounter about 300 million years ago and, although the Whirlpool escaped unharmed, it pulled many stars away from the smaller system, forming a faint bridge between them.

M106 This is a tilted spiral galaxy, shining with overall magnitude 9.1. It is relatively large for a spiral galaxy and shows hints of structure in even a small telescope.

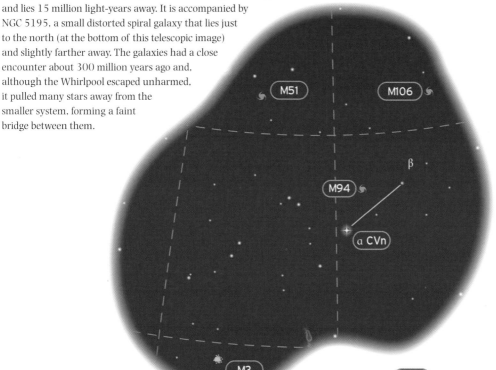

M94 This is another face-on spiral, roughly the same distance from Earth as the Whirlpool, and slightly fainter at magnitude 8.9. It is easily detectable through a small telescope, and has a tighter spiral structure than M51.

α CVn The constellation's brightest star, Alpha Canum Venaticorum, is named Cor Caroli, or Charles' Heart, after the executed British King, Charles I. Cor Caroli is a double white star of magnitudes 2.9 and 5.6, easily split with binoculars or small telescopes, and lies 82 light-years from Earth.

M3 The globular cluster M3 is a dense group of stars on the limit of naked-eye visibility at magnitude 6.2. M3 is 34,000 light-years away and contains about a million stars. Binoculars will show it as a fuzzy star, but a small telescope will reveal it as a mottled ball of light, a quarter of a degree across.

Clusters and superclusters

Galaxy clusters and superclusters are the largest structures in the Universe. Stretching across hundreds of millions of light-years, large clusters contain thousands of galaxies, while superclusters grow even bigger and have millions of members.

As a general rule, galaxies are all rushing away from one another as the Universe expands. But because they are not evenly distributed across space, they often tend to be drawn in one direction or another by the pull of their nearby neighbours. In this way, the general tendency to expand is overcome on a relatively small scale and various groupings of galaxies result.

Clusters and superclusters are collections of galaxies moving through space in the same direction, but gradually falling together under their own gravity. A cluster may be described as poor or rich depending on the number of galaxies in it. Poor clusters are also known as groups, our galaxy being in the Local Group. A further classification describes clusters as regular or irregular. Our Local Group

is irregular in shape while the Coma Cluster – being more spherical – is regular. A visible grouping of clusters is called a supercluster and may be up to 100 million light-years across.

Our Local Group is an outlying member of the Local Supercluster, centred on a huge, 3,000-member galaxy cluster in the constellation Virgo, about 50 million light-years away. The whole Local Group is currently falling toward the centre of the supercluster at 250 km/s (160 mps).

Large-scale maps of the position of millions of galaxies have shown us that they are not scattered evenly across the Universe. Clusters and superclusters seem to concentrate in narrow strips and sheets called filaments, on the edge of vast bubbles or voids of apparently empty space. The voids also appear to

be loosely similar in shape. Although this structure was apparently built into the Universe from its beginning, clusters and groups are still forming and evolving.

Astronomers can guess the age of a cluster by the number of galaxy collisions and mergers that have happened within it. Although this method is extremely conjectural, it seems that clusters broadly fall into two main types. Younger clusters contain mostly spiral galaxies, and have no obvious structure; older clusters contain more elliptical and lenticular galaxies, and have regular shapes. They may also contain one or more giant elliptical galaxies at their centres.

If a galaxy cluster contains giant ellipticals, then it has probably been evolving for several billion years – long enough for spiral galaxies to collide and coalesce into ellipticals, and for some of these ellipticals to sink toward the centre of the cluster and grow bigger by cannibalizing others.

Each new galaxy collision ejects hot, X-ray-emitting gas into intergalactic space. As other galaxies plough through this gas, they experience ram pressure, which strips away their gas without a direct collision. This is how lenticular galaxies – spiral galaxies whose star formation has been stopped – are thought to form. The theory is supported by the fact that lenticulars are more common in old and highly evolved clusters.

The distribution of hot gas in a cluster helps astronomers to estimate its age. As the cluster evolves, this gas will naturally tend to sink toward the centre, often collecting around the giant ellipticals. By studying the distribution of the gas in X-ray, astronomers can work out the cluster's history – for example, a recently swallowed group of galaxies may inject a separate blob of gas away from the centre.

The Virgo Supercluster This gigantic grouping of clusters, otherwise known as the Local Supercluster, forms a belt around the northern and southern heavens with the Virgo Cluster – a vast group of thousands of galaxies – at the centre.

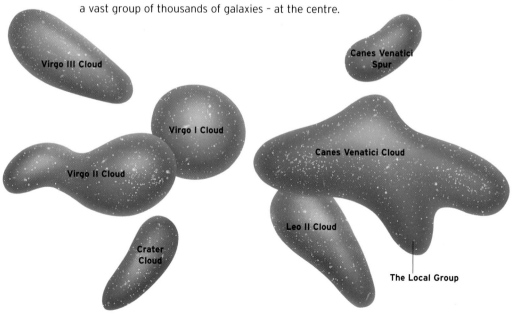

Virgo III Cloud

Canes Venatici Spur

Virgo I Cloud

Canes Venatici Cloud

Virgo II Cloud

Crater Cloud

Leo II Cloud

The Local Group

See also: The Local Group 144 · Colliding galaxies 152 · Structure of the Universe 164 · First stars & galaxies 176

Berenices' galaxies

12 **COMA BERENICES** *Sandwiched between Boötes and Leo, Coma Berenices, or Berenice's Hair, is a small but distinctive constellation. Although its stars are mostly faint, they are numerous – the constellation contains Melotte 111, one of the nearest open star clusters to Earth. The cluster covers several degrees of sky and is a beautiful sight through binoculars. Apart from its profusion of stars, Coma contains a rich concentration of galaxies, most belonging to the Virgo Supercluster. Beyond these relatively nearby galaxies, and beyond the reach of amateur telescopes, Coma also contains one of our nearest neighbouring superclusters.*

M64 Messier 64 is known as the Black-eye Galaxy because of its distinctive appearance, caused by a thick lane of dust silhouetted against the brighter central regions. At magnitude 9.3, M64 is easily seen in a small telescope, and under good conditions the dark dust patch can be detected with the smallest instruments. This galaxy lies roughly 30 million light-years from Earth.

M53 A medium-sized globular cluster shining at magnitude 7.6, M53 lies 56,000 light-years from Earth. Through binoculars it appears as a fuzzy star half the size of the Full Moon, but a small telescope will resolve the individual outer stars. To its southeast lies the much sparser NGC 5053.

NGC 5053 This is an unusually sparse globular cluster, containing less than 4,000 stars, and shining at magnitude 9.5. Its comparatively large size means it is very dim, and only visible when using medium-sized telescopes.

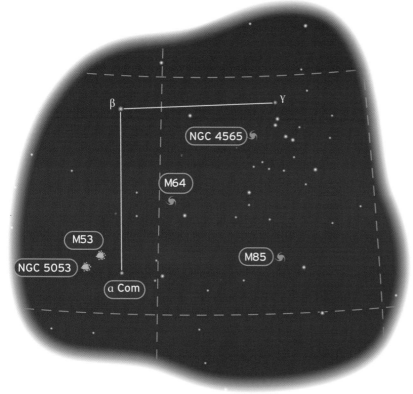

M85 One of the sky's brightest lenticular galaxies, M85 is a small, magnitude 10.0 disc of light, visible in small telescopes as an oval glow with a condensed central nucleus. It lies 40 million light-years away.

α Com Alpha Comae Berenices is a close binary system: two yellow stars of magnitude 5.1 shining at a combined magnitude 4.3. The stars, about 47 light-years from Earth, orbit each other every 26 years and are barely divisible even in medium-sized telescopes.

NGC 4565 This is the sky's brightest and most elegant edge-on spiral. Through small telescopes, it appears as a sliver of light a quarter of a degree long, with a bright spot in the centre marking its nucleus. The galaxy is about 30 million light-years away and shines at magnitude 10.3.

156 Active galaxies

Some galaxies show activity that cannot be explained in terms of even the most brilliant starbursts or violent collisions. Such systems may have bright central nuclei, or emit intense radio waves from giant clouds above and below them.

Active galaxies were discovered in the 1950s, when the first large radio telescopes were built. Astronomers found that some of the largest and most intense radio sources in the sky were huge clouds with no visible counterparts. However, many of these lobes were found in matching pairs above and below galaxies – particularly ellipticals.

The lobes emit radio waves by a specific method called synchrotron radiation, which produces a distinct range of low-frequency radio waves. This is the signature of electrically charged particles spinning through an intense magnetic field, far stronger than our own galaxy's field. As radio astronomy developed, astronomers detected long, straight jets linking the central galaxies to the lobes, and found hot spots on the outer edges of the lobes. These regions of intense radiation form where the ejected particles slow down and pile up as they plough through the thin material of intergalactic space.

So radio galaxies appear to have enormous magnetism and a power source that constantly pumps charged particles into their jets and lobes. Why should they behave so differently from normal ellipticals? One clue comes from their appearance: they often show unusual amounts of dust and distorted appearances, which indicate that they may have recently collided with and absorbed other galaxies.

Another type of activity seems less violent and occurs in spiral galaxies. These "Seyfert galaxies" have unusually bright cores that emit radio waves, ultraviolet and even X-rays, indicating they are extremely powerful energy sources. Like radio galaxies, Seyferts often seem to have been involved in galactic collisions or close encounters.

Although the source of a Seyfert's energy is hidden by stars and dust in visible light, X-rays and radio waves can leak through the dust to reach us. They show that the power source is concentrated in a tiny region at the galaxy's nucleus – a brilliant disc of material just a few light-years across. It seems that something at the heart of the galaxy is creating this disc and heating it intensely. Infrared images of the central regions in radio galaxies show that these also contain a hot central disc of material that seems to be the source of their jets.

Astronomers think that the activity in these galaxies is caused by a central black hole with a mass of millions of Suns. Such black holes are thought to lie at the centre of most galaxies, including our own Milky Way. As active galaxies do not maintain the same level of activity forever, but settle down eventually, it may be that currently inactive galaxies were once active.

The black holes are believed to form when collapsing gas clouds in the nucleus run away with themselves during the galaxy's formation. But to fully understand the mechanism behind active galaxies, we need the final piece of the jigsaw: extremely violent galaxies at the edge of space called quasars.

A radio galaxy The elliptical galaxy Centaurus A is one of the most powerful radio emitters in the universe. It is seen here crossed by its characteristic band of light-obscuring dust.

See also: Black holes 126 · Types of galaxy 148 · Elliptical galaxies 150 · Quasars & blazars 158

Radio transmitters in the night sky

46 CETUS *Although Cetus means "Whale", the constellation that bears this name was traditionally seen as the sea monster that threatened Andromeda in the Perseus legend. To the naked eye it seems little more than a dull collection of stars straddling the celestial equator, and rises highest in the sky on evenings around October. However, since it is well away from the Milky Way, Cetus offers clear views of several interesting galaxies, including a nearby Seyfert galaxy. It also contains some very interesting stars.*

o Cet One of the sky's most famous variable stars, Omicron Ceti, or Mira, varies between magnitude 2.0 and magnitude 10.1 in a cycle of about 332 days. Its changes are easy to follow in binoculars or a small telescope, and sometimes it can reach 2nd magnitude. First recognized in 1596, Mira is the prototype long-period variable, a slowly pulsating red-giant star. It lies 420 light-years away.

M77 Messier 77 is one of the closest Seyfert galaxies, making it a magnitude 9.7 object even though it is about 50 million light-years away. Through a small telescope it appears as a small disc of light as the spiral lies face on to Earth. Larger instruments reveal that the source of the galaxy's light is an abnormally bright nucleus.

τ Cet Tau Ceti is one of the nearest stars to the Sun, just 11.9 light-years away. It is a yellow, main-sequence star similar to the Sun in many ways, and shines at magnitude 3.5. Tau Ceti is a possible haven for alien life relatively close to Earth, although as yet there is no evidence to suggest it even has planets orbiting it.

UV Cet UV Ceti is a faint star system even closer to Earth than Tau Ceti, at 8.4 light-years. The two stars are red dwarfs with about 8 per cent of the Sun's mass—about as small as a star can get. They shine at magnitudes 12.4 and 13.0. The fainter star is a flare star, wracked by stellar flares hundreds of times more powerful than the Sun's, which occasionally cause it to brighten.

OTHER ACTIVE GALAXIES

NAME	CONSTELLATION	DISTANCE	TYPE	MAGNITUDE
3C 405 (Cygnus A)	Cygnus	600 million light-years	Radio galaxy	18.0
NGC 253	Sculptor	9 million light-years	Seyfert galaxy	8.0
NGC 1275 (Perseus A)	Perseus	230 million light-years	Seyfert/radio galaxy	13.0
NGC 1566	Dorado	50 million light-years	Seyfert galaxy	10.2
NGC 4261	Virgo	100 million light-years	Radio galaxy	11.4
NGC 5128 (Centaurus A)	Centaurus	15 million light-years	Radio galaxy	7.0

158 Quasars and blazars

Quasars are the most violent objects in the Universe – starlike points of light that are, in fact, intense sources of radiation at the edge of the observable Universe. They are so far away that their light has taken billions of years to reach us.

When astronomers discovered quasars in the 1950s, they thought they were relatively nearby. The name is short for quasi-stellar radio sources, for this is what they are: objects that appear like stars in visible light, but are also strong sources of radio waves.

As most radio sources in the sky have a perceptible size, these pointlike radio sources were a puzzle, and when their spectra were analysed, they showed bright emission lines that could not be traced to any known element. In 1963, Maarten Schmidt of the Mount Palomar Observatory realized they were the normal lines formed when hydrogen emits light, but shifted by an enormous amount toward the red end of the spectrum. This redshift was the largest ever found, and meant that the quasar must be hurtling away from Earth at tremendous speed. Other quasars were soon found with even higher redshifts. It is now accepted that quasars are extremely distant objects, on the edge of the cosmos, and that they formed billions of years ago, in the early days of the Universe.

A different type of quasarlike object was discovered in 1968, when a rapidly variable star called BL Lacertae turned out to also have a high redshift and to produce radio signals. This was the prototype BL Lac object, or blazar. Because its light varied from day to day by up to 100 times, the blazar had to be generating all its energy in a very small region, barely bigger than our Solar System. If the power source was any larger, then even if its fluctuations travelled at the speed of light, they could not affect the entire structure in a day.

It seems that a single theory can explain both quasars, blazars and other active galaxies. This model is based on the growing evidence that most galaxies have a supermassive black hole at their centre. In normal galaxies this black hole may have long since cleared out the region around it, and now lies dormant. Occasionally, when a rogue star cluster comes too close, the sleeping giant wakes. Its enormous gravity shreds anything within reach, pulling it down into a spiral accretion disc before it disappears into the black hole.

The size of the black hole and the amount of material poured into it can create different levels of activity. In Seyfert galaxies, the activity is low and merely causes an unusual brightening of the core. In more violent types, the accretion disc is heated so much that it emits X-rays, and radio jets are ejected from the black hole. We see radio galaxies when we see the galaxy edge-on, and its central disc is hidden from view. In quasars, we can see the X-ray emitting disk as well as the radio jets, whereas in blazars we are looking straight down the galaxy's radio jet and into the rapidly changing accretion disc.

An active galaxy's age also affects its activity – quasars and blazars are young galaxies and so far more material seems to be pouring into the black hole.

A typical active galaxy with (inset) the black hole at its centre At left, the jet of particles that create the galaxy's radio lobes can be seen pouring out of the centre. As material of all kinds is pulled into the black hole, enormous quantities of energy are given off. Quasars are a very special and violent type of active galaxy, similar in some ways to radio and Seyfert galaxies closer to us. This was confirmed when radio jets were found blasting out from some quasars, expanding into lobes in interstellar space.

See also: Types of galaxy 148 · Active galaxies 156 · First stars & galaxies 176 · The invisible Universe 182

Exploring Sculptor

NGC 253 This spiral galaxy lies edge-on to Earth. Although it is 9 million light-years away, it is visible through small telescopes or binoculars. Shining at magnitude 8.0, NGC 253 looks like a cigar-shaped cloud of light, as long as the Full Moon is wide, and overlaid by dark dust lanes. The nucleus, where an active black hole probably lurks, is a bright starlike point in the middle of the nebulosity.

80 **SCULPTOR** *Lying to the south of Cetus and Aquarius, the small and faint constellation of Sculptor has little to attract the attention at first sight – the brightest star is magnitude 4.4. However, Sculptor is home to one of the nearest neighbouring groups of galaxies, dominated by the large spiral NGC 253. This galaxy, with its active nucleus generating X-rays, gamma rays and unusual amounts of light, is like a quasar in miniature. The constellation also contains many examples of the real thing, although these are beyond the reach of amateur telescopes.*

NGC 55 NGC 55 is a slightly smaller and fainter edge-on spiral than NGC 253, at magnitude 8.1. It is also a member of the nearby Sculptor Group, roughly 7 million light-years from Earth, and should be visible through binoculars under dark skies.

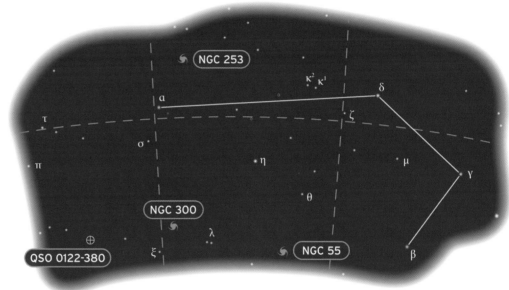

NGC 300 The spiral galaxy NGC 300 lies face on to Earth, but it is hard to detect anything more than its central regions with a small telescope. Its overall brightness is magnitude 8.8, and it lies about 8 million light-years from Earth.

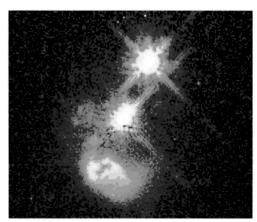

QSO 0122-380 Sculptor contains several quasars, but most are beyond the reach of small telescopes. QSO (Quasi-Stellar Object) 0122-380 is a relatively bright one – at magnitude 16.5 it is a little fainter than Pluto. It has a redshift of 2.2, indicating that it is moving away from us at around 80 per cent of the speed of light. This Hubble Space Telescope image shows the heart of a quasar fuelled by a galactic collision.

The edge of infinity In the early days of the cosmos, a quasar galaxy blasts its brilliant radiation out toward the far corners of the Universe, where, 10 billion years later, its signal will finally reach Earth. At the quasar's heart lies a monstrous black hole swallowing everything within its reach. As the black hole rips whole stars apart, it surrounds itself with a brilliant disc of superheated matter that generates more radiation than the rest of the galaxy put together.

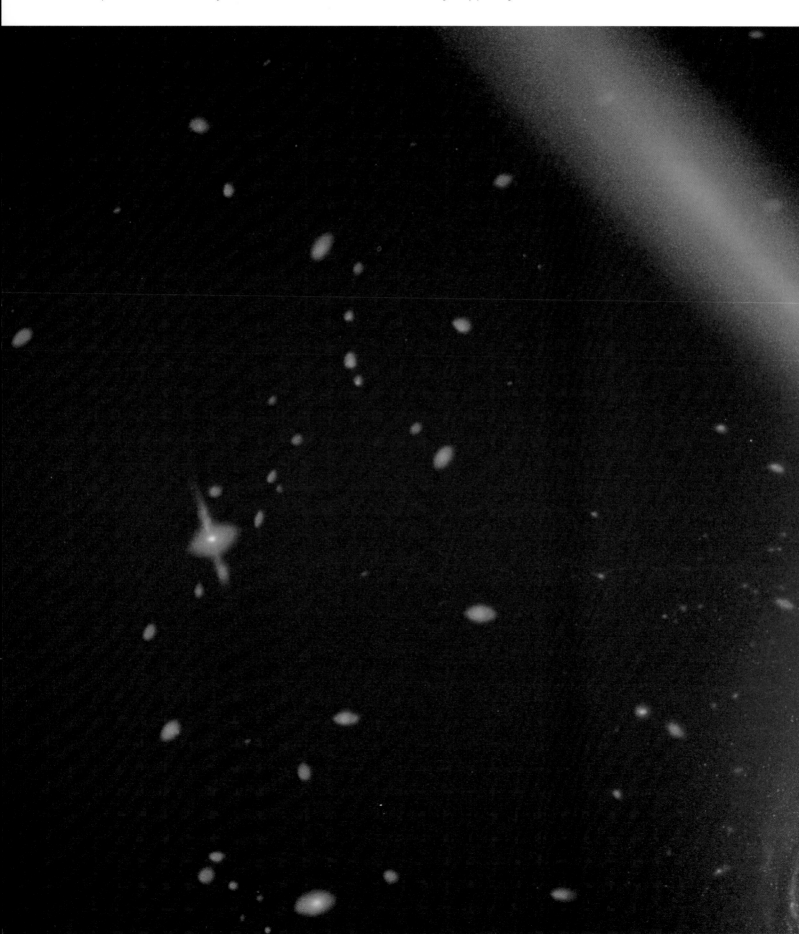

Foundations

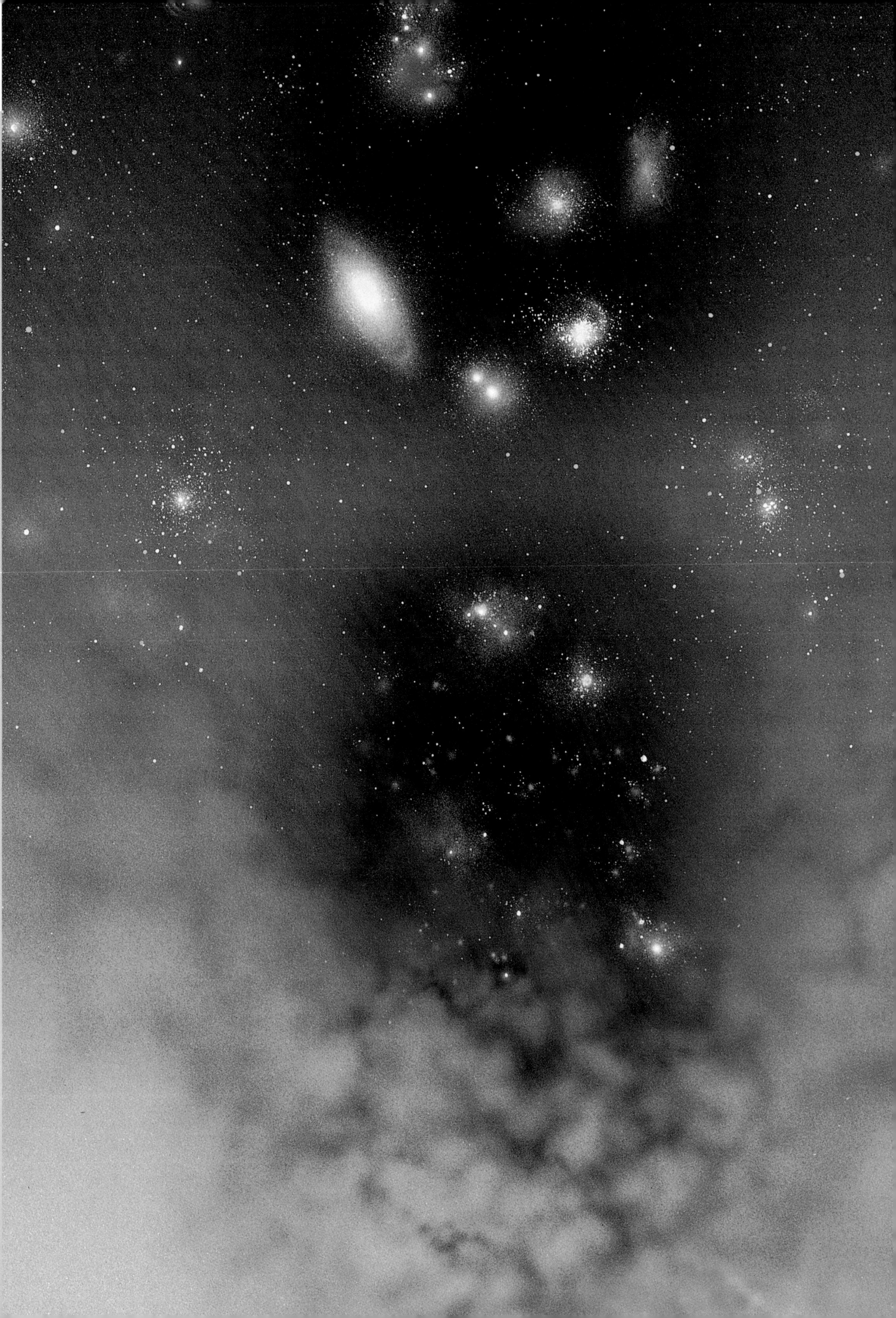

Introducing cosmology

Beyond the stars, galaxies, clusters and superclusters, some astronomers seek the answers to fundamental questions about the Universe: its structure, origins and fate. Once the province of theologians and mythmakers, only in the last century has cosmology blossomed into a true science.

People have pondered the nature of the Universe for thousands of years, looking to the stars to explain not only events on Earth, but also their place in a wider cosmos. However, it is only in the past few decades that science has approached a true understanding of the nature of the Universe. Until early in the 20th century, it seemed that our Milky Way Galaxy was all there was. Few astronomers suspected that the smudgelike "nebulae" which dotted the sky were huge stellar systems far beyond our own. The rapid advances made since then have given rise to the many new and surprising ideas that are covered in the following pages. Though many of these may seem counter to common sense and everyday experience, they are theories derived from, and confirmed by, our observations and experience of the Universe – hopefully they lead us closer to a true understanding of our place in the cosmos.

Modern cosmology has been shaped by changing technology – new and better telescopes, radio astronomy and orbiting observatories. Great leaps forward in theoretical physics, including speculations about the nature of space and time, and the birth of atomic physics, have also aided the understanding of otherwise inexplicable celestial phenomena. Although cosmologists rely on information gathered by observational astronomers, they are themselves more likely to be found hunched over complex mathematical formulae. Only by unifying the findings of many branches of physics and astronomy can they hope to reveal the secrets of the cosmos in which we live.

BEGINNINGS AND ENDINGS

The discovery of galaxies beyond the Milky Way revealed for the first time that the Universe is expanding away from us in every direction. This in turn gave rise to the Big Bang theory of creation, the jewel in the crown of modern cosmology. Because everything in the Universe is moving apart, there was clearly a time in the distant past when things were much closer together. Taken to its logical conclusion, all of the Universe must have originated at a single point in space and time: an immense explosion called the Big Bang.

One of the greatest challenges of modern cosmology has been to estimate how much the Universe has grown since its birth, precisely how old it is and how far away its most distant objects are. The Hubble Space Telescope was specifically designed to address this question by accurately

measuring the distances of the farthest objects in the Universe.

Cosmology is concerned with our destiny as well as our origins: how will the Universe end? The answer depends on the rate at which the cosmos is expanding and the precise amount of material it contains, which, in turn, depends on measuring the potentially huge amounts of invisible dark matter in the Universe. Ultimately, there are two possible scenarios. If the Universe has enough mass, gravity will eventually slow its expansion until it reverses, collapsing inward to a "Big Crunch". If there is too little mass, the Universe will expand forever, until the last stars exhaust their fuel and die.

Cosmic detectives
Cosmologists use facts about the Universe to build a story of how it formed and evolved into the place we see today.

Kaleidoscope galaxies
This Hubble Space Telescope photograph shows gravitational lensing, one of the effects predicted by relativity. The blue objects are multiple images of a distant galaxy produced as its light is bent by gravity around a much nearer, yellow galaxy cluster.

The largest objects in the Universe are the clusters and superclusters of galaxies. By studying patterns in their distribution, astronomers can learn much about the overall structure of the cosmos and our position in it.

According to Hubble's law, the more distant a galaxy is, the faster it is moving away. Most large-scale maps of the Universe are based on this law, using the redshifts of galaxies as a direct measure of their distance.

Huge conglomerations
Clusters and superclusters of galaxies resemble huge bubbles that span millions of light-years in this map of the nearby Universe. The Local Supercluster contains thousands of bright spiral and elliptical galaxies, as well as countless faint dwarfs.

SUPERCLUSTERS

The Milky Way and the nearby Local Group of galaxies are just a small cluster on the outskirts of the vast Virgo Supercluster. This supercluster is one of many stretching across vast expanses of space which, unlike individual galaxies and smaller clusters, are not isolated from one another. Superclusters are created as galaxies gather around regions of higher gravity and fall into orbit around each other.

Astronomers map the location of galaxies relative to the Milky Way by measuring the redshift of their light. They assume that the redshifts are caused by the Doppler effect – a decrease in the frequency of waves reaching an observer when the source of those waves is moving away. Thus, redshifts are a measure of the rate at which a galaxy is receding from us.

GALACTIC DECEPTIONS

Charts of our region of the cosmos show several interesting features. One is that most galaxy clusters seem to be stretched into ovals radiating away from the Local Group. This effect arises because Hubble's law is true only for the Universe in general. On the smaller scale of galaxy clusters or even superclusters, it collapses.

Although a cluster's overall motion follows the expansion of the Universe, the galaxies within it can move in any direction as they fall toward one another—for example, the Milky Way and the Andromeda Galaxy are closing in on each other at 130 km per second (80 mps). Individual galaxies within a cluster can therefore show a greater or smaller redshift than the cluster average. There is no way to separate the spectral shifts caused by individual galaxy

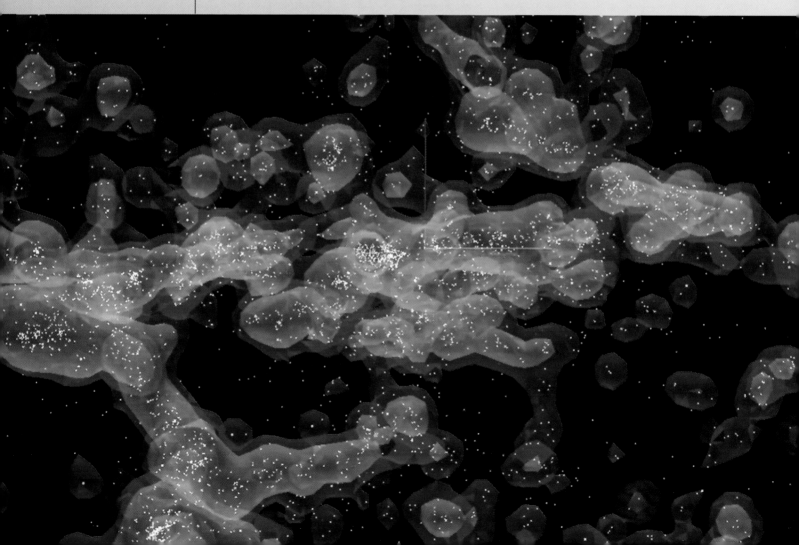

movements from that of universal expansion, so the clusters unavoidably appear warped.

On a larger scale, even superclusters can be diverted from the general expansion of the Universe by objects with sufficient gravity. The Virgo Supercluster is being pulled toward a massive object called the Great Attractor. Lying some 150 million light-years from Earth, it is pulling in everything within a space of 300 million light-years. The Great Attractor seems to be a super-supercluster, with a very dense concentration of mass at its centre.

GALACTIC BUBBLES

By far the most noticeable features on maps of the local Universe are the large empty bubbles in the distribution of galaxies. Clusters and superclusters appear in the gaps between these voids, and extended sheets and filaments of galaxies stretch around their edges, linking the superclusters together.

These filaments and voids are the largest structures in the Universe; some of them are more than 300 million light-years across. They contain no visible matter but this does not necessarily mean they are entirely empty. One suggestion is that they may be filled with billions of dwarf galaxies, too dim to see with even the most powerful Earth-based telescopes. They could also contain large amounts of cold hydrogen and helium left over from the creation of the Universe. By studying the spectrum of light from the most distant galaxies, astronomers have discovered a forest of dark lines where intergalactic hydrogen clouds with different redshifts are absorbing light. This suggests that there may still be shreds of gas in the space between galaxy clusters.

DARK MATTER

If there is matter in the voids, it could have an important effect on the fate of the Universe. Although the cosmos is currently expanding, its mass might one day be enough to slow the expansion and even cause it to reverse. Both the Big Bang theory of creation and models of the way galaxies rotate indicate that visible matter accounts for only about 10 per cent of all the mass in the Universe. The remaining 90 per cent must be dark matter but the precise mass and nature of the unseen material is unknown. Some galaxies seem to have vast halos of matter

extending far beyond their visible limits. Such halos might be made up of "MACHOs" – Massive Compact Halo Objects – that include near-invisible brown dwarfs, black holes and dead stars. Between the galaxies, space might also be flooded with heavy but undetectable "WIMPs" – Weakly Interactive Massive Particles.

UNIVERSAL INFINITY

It is important to realize that the Universe we can see from Earth is just a small part of the overall cosmos. The observable Universe stretches for 12 billion light-years in every direction – the maximum distance light could have travelled since the Big Bang. As we look through space, we are also looking back through time until, at the edge of the observable Universe, we are seeing back to the earliest days of the cosmos. But at this

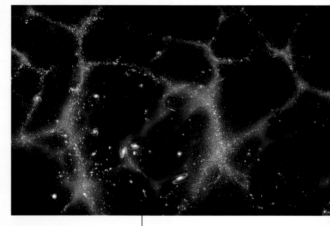

instant, an observer on a planet at the edge of our observable Universe can see for a further 12 billion light-years. They may be able to see the first galaxies forming in our region of space, while also seeing another 12 billion light-years in the opposite direction.

So does the Universe stretch away like this forever? The answer depends again on the amount of mass in the Universe. According to Einstein's general theory of relativity, large masses can bend space around them. If the Universe contains enough mass, it might bend space back on itself through a fourth dimension. One way to visualize this warping of the cosmos is to see three-dimensional space as the two-dimensional surface of an expanding balloon. A circle around any point on the balloon might mark the limit of that point's observable Universe and, if the balloon has expanded far enough, then space within that circle will appear to be flat. But a straight line can be drawn that continues far beyond the circle. From any point on this line, the Universe would also appear to be flat but, eventually, the line would return to the point from which it started.

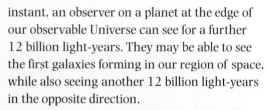

Filaments and voids
Seen from an imaginary viewpoint in deep space, galaxy filaments mark out huge, apparently empty, voids hundreds of millions of light-years across.

The expanding cosmos

Perhaps the single most important fact about the Universe is that it is expanding. The farther away we look in space, the faster objects are receding from us. This apparently simple discovery has profound implications for the past, present and future of the Universe.

HUBBLE'S LAW

The expansion of the Universe was first discovered by Edwin Hubble, the US astronomer who proved that other galaxies lay far beyond the Milky Way. He found that distinctive spectral lines in the light from other galaxies were shifted away from their expected positions toward the low-frequency, red end of the spectrum. Most galaxies beyond our immediate Local Group showed such redshifts – probably, Hubble suggested, caused by the Doppler effect. In other words, the galaxies were moving away from us.

Hubble soon found that the more distant a galaxy is, the greater its redshift and the faster it is receding from us. This discovery, called Hubble's law, only makes sense if every point in space is moving away from its neighbours, and the Universe as a whole is expanding. Hubble showed that the speed of a galaxy's recession depends on its distance from us multiplied by a simple constant. This factor, called the Hubble constant, H_0 (pronounced "aitch-nought"), is commonly called the rate of expansion of the Universe. H_0 is measured in kilometres per second per megaparsec – a megaparsec (Mpc) is 3.26 million light-years – so it actually measures how the speed of recession increases with distance. This is what makes H_0 a description of expansion, not acceleration.

Hubble's law allows astronomers to use redshift as a measure of the distance to faraway objects. But they still need to know the precise value of H_0. This is difficult as there is no way of independently measuring the distance to galaxies beyond our immediate locality.

For many years, estimates of H_0 ranged from 50 to 100 km per second (30 to 60 mps) per Mpc. In other words, for every 3.26 million light-years farther away we look, galaxies will, on average, recede 50 to 100 km per second (30 to 60 mps) faster. Since its launch, the Hubble Space Telescope has conducted a detailed programme to measure the distances of faraway galaxies using standard candles – objects such as Cepheid variables whose true brightness can be easily worked out. As a result, the value of H_0 has closed in on 70 km/s/Mpc (44 mps/Mpc).

DOUBTS ABOUT EXPANSION

The entire model of a rapidly expanding Universe is based on the assumption that

Rates of recession
If space were the surface of an expanding balloon, Hubble's law would still apply. As the balloon grows bigger, every inch of its surface stretches at the same rate. So the distance between faraway galaxies (measured across the surface) grows rapidly, while the distance between neighbouring galaxies grows more slowly.

Hubble's law is correct and that the redshifts of distant galaxies are caused by the Doppler effect. This is accepted among most astronomers but there are still some with doubts. One, for example, claims to have found physical links, such as chains of stars, linking objects with wildly different redshifts. If galactic redshifts are not caused by their movement, then what could produce them? One suggestion is that light waves simply lose energy and become redder as they cross large distances in space.

THE AGE OF THE UNIVERSE

The Hubble constant can provide very useful information about the Universe. For example, if it is expanding outward, it must have originated at a single point. Furthermore, assuming that the Universe has expanded at the same rate ever since the Big Bang, the distance to the farthest objects, at the edge of our observable Universe, will be the same in light-years as the age of the Universe is in years, because light from these objects has taken all of that time to reach us.

Finding the age of the cosmos in this way has problems, however. Since our estimates of the distance to the farthest objects are dependent on Hubble's law, they will be affected by the accuracy of the Hubble constant. A large Hubble constant means a fast expansion and a young age for the Universe, whereas a low constant produces a slow expansion, which means a greater age. According to the best modern values for H_0, the Universe is approximately 12 billion years old, with a 10 per cent margin of error either way.

But there is another potential problem. The expansion of the Universe is not constant. The gravity of all the matter it contains should be slowing it down, perhaps imperceptibly. The rate of this slowdown will determine the fate of the Universe. If expansion is slowing rapidly, gravity might pull the Universe back into a Big Crunch. If it is slowing more gradually, then the expansion may simply continue forever.

In the late 1990s, astronomers set out to settle the issue once and for all by finding another way to determine the distance of the farthest galaxies. They conducted an intensive search for white dwarfs collapsing into neutron stars (so-called Type Ia supernovae) in quasar galaxies billions of light-years away and billions of years back in time. Such supernovae involve a

On the increase
This Hubble Space Telescope photograph shows a supernova in a galaxy several billion light-years from Earth. Explosions such as this indicate that the Universe's rate of expansion may in fact be increasing.

process that always releases the same amount of energy with the same peak luminosity. By measuring their brightnesses, it is relatively easy to work out how far away they are.

What the supernova cosmology project found surprised almost everyone. If the Universe was slowing down, then the quasar distances found from supernovae would be smaller than those derived from redshifts alone. But time after time, the quasars turned out to be farther away than expected. This surprising discovery means that the rate of expansion is accelerating as it gets older, dooming the cosmos to expand forever, gradually becoming darker and colder as stars and galaxies exhaust all their available fuel.

But how can the expansion be accelerating? One answer may lie in Einstein's theory of gravity. Working a decade before Hubble's discovery of the galactic redshifts, Einstein believed that the Universe was static, so he thought there must be some outward force preventing everything from collapsing under gravity. He therefore introduced a cosmological constant into his calculations, adjusting his theory to fit contemporary thinking. Einstein later dismissed the cosmological constant as his biggest blunder but it now seems he may have been right all along. While the acceleration is still controversial, theoretical physicists are working on a new version of the cosmological constant, called vacuum or dark energy.

So near, yet so far Despite discovering the expansion of the Universe, Hubble was not entirely right. By his calculations, the Milky Way was the biggest of all galaxies and the Universe was younger than the actual age of Earth. Since then, the value of his constant has been recalculated, and a more accurate age for the Universe found.

The nature of space-time

Einstein's theory of relativity is the most important tool for today's cosmologists. Although Isaac Newton's simple, elegant theories of gravity and mechanics still work at small speeds and distances, relativity can describe how the Universe works on the grand scale.

THE MEANING OF RELATIVITY

Relativity is based on one simple observation: that the speed of light is apparently constant throughout the Universe, regardless of how an observer is moving. In other words, it does not matter whether you are moving toward or away from a source of light waves; they will always reach you at the same speed, which is around 300,000 kilometres per second (186,000 mps).

This curious fact robs us of any way to measure our absolute position in the Universe: if light did not behave in this way, it would be possible to find out how we were moving through space by measuring the variations in speed of light coming from different directions. Einstein realized the implications of the unchanging speed of light – the Universe will look the same to any observer in a situation where they are not accelerating or decelerating.

Why should light behave in this way? The answer lies in its unique nature, which only became apparent in the early 20th century. Light can behave both as a wave and as a particle called a photon. Photons carry energy locked up within them, but they are massless particles and are therefore freed of many physical constraints. This allows them to travel at the Universe's ultimate speed limit – a speed that objects with mass can approach but never reach.

The behaviour of light defies common sense, and Einstein realized that physics would have to be rewritten to accommodate it. Fundamentally, he suggested that time was a fourth dimension that could be treated in the same way as the three dimensions of space, and that all four were intrinsically linked together in a space-time continuum. He published his theories in two parts: special relativity in 1905, and general relativity in 1915.

TRAVEL NEAR THE SPEED OF LIGHT

Special relativity explains how objects behave when travelling close to the speed of light. The dimensions of an object that we perceive are the visible three-dimensional part of a four-dimensional property. According to Einstein, objects have a four-dimensional equivalent of length called extension. As an object approaches the speed of light, our perspective changes, so that we see less of its length and more of its extension. The object appears to shrink along the direction of travel, while time appears to stretch and slow down for it.

This remarkable effect, called time dilation, is imperceptible in most everyday situations because the speeds at which objects move are only a tiny fraction of the speed of light. However, scientists have flown atomic clocks on jet aircraft and found, after landing, that they have lost trillionths of a second. An important point is that the effect is relative: an astronaut

Distorted space-time

According to general relativity, a massive object warps space-time around it. In three dimensions, this is best shown by the hourglass shape (near right). But it is often easier to use a two-dimensional analogy, with space as a flexible sheet (far right). Any massive object creates a dent in the sheet, called a gravitational well, and objects naturally roll in toward it.

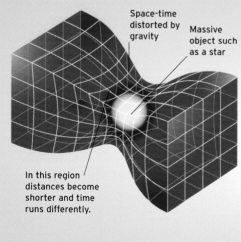

Space-time distorted by gravity

Massive object such as a star

In this region distances become shorter and time runs differently.

Distortion in 3-D

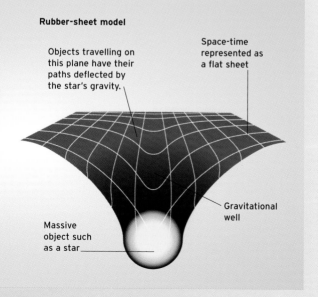

Rubber-sheet model

Objects travelling on this plane have their paths deflected by the star's gravity.

Space-time represented as a flat sheet

Gravitational well

Massive object such as a star

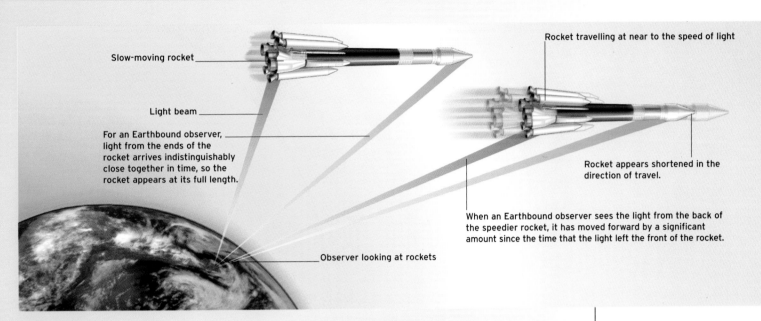

Slow-moving rocket

Light beam

For an Earthbound observer, light from the ends of the rocket arrives indistinguishably close together in time, so the rocket appears at its full length.

Observer looking at rockets

Rocket travelling at near to the speed of light

Rocket appears shortened in the direction of travel.

When an Earthbound observer sees the light from the back of the speedier rocket, it has moved forward by a significant amount since the time that the light left the front of the rocket.

travelling at the speed of light would not be aware of the time dilation visible to an observer on a nearby planet. Indeed, for the astronaut, it would appear that it was the observer whose time was slowing down.

RELATIVITY AND GRAVITATION

Einstein's General Theory of Relativity explains the true nature of the space-time continuum in which objects exist. Most importantly, it explains the origin of gravity – the force that dominates the Universe on the largest scale.

According to Einstein, space is flexible. Large concentrations of mass, such as stars and planets, pinch space inward around them, causing it to balloon out in the fourth dimension of time. In three dimensions, space appears to form an hourglass shape around a massive object, but it is often simpler to think of space as a two-dimensional rubber sheet. Heavy masses create dents or gravitational wells in the sheet of space, which affect other objects passing nearby. For example, the planets of our Solar System are trapped in orbit around the Sun by the gravitational incline toward its centre, though they are travelling fast enough to keep moving around the edges of the well, and not fall into the centre.

As with special relativity, general relativity has been proven by experiment. During eclipses, astronomers have observed that stars close to the edge of the Sun appear to have shifted position. In fact, light from the stars has bent as it passes through the Sun's gravitational well.

This effect, called gravitational lensing, has also been seen by the Hubble Space Telescope in deep space, where the gravity of galaxy clusters bends and focuses light from objects far beyond them, just like a spectacle lens.

One of relativity's main predictions is the existence of gravitational waves – ripples in space-time spreading out from regions where a large gravitational well is rapidly changing. For example, binary pulsars – two neutron stars orbiting close to one another – should be strong generators of gravitational waves. The passage of a gravitational wave should cause a minute change in the shape of space within it, and physicists are now working on several major projects to detect these changes.

General relativity also provides a new explanation of black holes, where objects entering them are falling into a deep, steep-sided pit in space-time. As they disappear out of our perceptible three dimensions, their extension becomes more visible and they experience time dilation. Einstein suggested that the bottom of a black hole might be capable of linking to a black hole in another part of the Universe. However, the bridge between the two holes would only exist for an instant, meaning that communication or travel between the two regions would be impossible. In the 1970s, cosmologist Kip Thorne suggested that it might be possible to stabilize this bridge and expand it to form a "wormhole". Science-fiction writers seized on this with glee and they would have you believe that every advanced civilization commutes by wormhole!

Time dilation Light travelling from the more distant end of a rocket has to leave an instant before light from the closer end, in order to reach an observer on Earth at the same time. A slow-moving rocket hardly moves in this time, but one travelling near the speed of light has moved forward by a significant amount. As a result, the two light sources are closer together and the rocket appears to be shorter. Light, however, still takes the same time to travel along the length of the rocket, so the speed of light, and also time, appear to have slowed down on board.

The moment of creation

A split second
Matter, energy, space and time were seemingly created out of nothing when the Big Bang occurred more than 12 billion years ago. An infinitesimal fraction of a second later, inflation blew up the infant cosmos to incredible size, creating vast amounts of energy that eventually formed particles, atoms, stars and galaxies.

The Big Bang theory is the crowning glory of modern cosmology – the best scientific model ever developed for the origins of the Universe . According to the theory, the Universe was born in an unbelievably powerful explosion around 12 billion years ago. In the first fraction of a second, all the matter in the cosmos was created.

IN THE BEGINNING

For the first instant – the first 10^{-43} seconds (see glossary for an explanation of some of the extreme numbers quoted in cosmology) – the Universe was so small that there were simply no rules. The laws of physics break down completely below a critical threshold known as an object's Compton wavelength, so the events that triggered the birth of the Universe remain mysteries that are, by definition, unsolvable.

This has not stopped astronomers speculating, however. Some believe that the Universe erupted spontaneously from an empty void, while others suggest it was born in the fiery collapse of a previous Universe from a Big Crunch. The renowned physicist Stephen Hawking has speculated that the Big Bang might have been triggered by a fourth dimension of space spontaneously transforming into time. Whatever triggered this cataclysm, it released huge amounts of energy and created space and

time as we know them. In some ways, it is meaningless to ask what happened before the Big Bang, or even to ask where it happened – space and time were born in that instant. Without time, space could not evolve. And without space, time could not move forward.

Precisely 10^{-43} seconds into the Big Bang, the Universe grew larger than its Compton wavelength, and the normal laws of physics began to operate. From this point onward, cosmologists are on firmer ground in their speculations. For a further fraction of a second, the whole Universe was far smaller than an atom, and quantum physics – the special set of rules that governs the subatomic realm – applied. Quantum theory allows particles to behave like waves, waves to act like particles, and mass (m) and energy (E) to interchange according to Einstein's famous equation $E = mc^2$ (with c being the speed of light). Similar conditions can only be recreated today in the most powerful particle accelerators.

BORN IN FIRE

Because the Universe was so compressed, its temperature was inconceivably high – around 10,000 trillion trillion degrees. Subatomic particles winked in and out of existence, including many unstable, high-mass, exotic particles that cannot exist in the low energies of our modern Universe. Some of these decayed into smaller, lower-energy particles such as electrons, quarks and neutrinos, which are still with us today.

For all its fury, the Big Bang was relatively small. The energy it released was equivalent to the mass of just a few pounds of material, and the infant Universe expanded reasonably slowly. If the universe had continued to expand at this rate, it would have been overcome by its own gravity and collapsed back on itself. But then something amazing and unimaginably violent happened – a small section of the Universe ballooned outward, blown apart by an enormous release of energy, to form the seed of the cosmos we know today.

COSMIC BLOW UP

This event, known as inflation, seems to have been triggered as the four fundamental forces of the modern Universe came into being. In the searing conditions of the Big Bang itself, these forces – gravitation, electromagnetism, and the weak and strong nuclear forces that bind subatomic particles were wrapped together in one superforce. Gravitation separated itself from the superforce instantly, but the other three forces remained bound together until the Universe reached 10^{-35} seconds old.

Then, the forces suddenly broke free of each other. This change of state was rather similar to the way water freezes: the formation of ice releases energy without changing the water's temperature. During inflation, the energy released as the fundamental forces froze out of the superforce powered the expansion of a small portion of the primordial Universe.

Inflation was over in an instant – by the time the Universe was 10^{-32} seconds old. But it left a very different Universe in its wake. Because it had expanded from a tiny region of space, the post-inflation Universe was very smooth and well mixed. The vast expansion also caused the temperature of the Universe to dwindle to almost

nothing – absolute zero or -273 °C (-460 °F) – before recovering, as it was flooded with more energy from the separating forces.

AFTERMATH OF INFLATION

There was now a comparatively enormous volume of space containing very little material but vast amounts of energy. In these conditions, $E = mc^2$ still applied, and much of the energy was swiftly converted into matter – matched pairs of particles and antiparticles (identical to normal particles, but with an opposite electric charge). In most cases, the particle pairs soon collided again, annihilating each other and releasing a burst of gamma rays. But a surge of energy at the end of inflation set particles and antiparticles free of each other. From this point on, as the temperature dropped rapidly, the creation of matter came to a swift end.

As the amount of free energy in the Universe dwindled, the mass of the particles that could be produced diminished. Exotic heavy particles, which can be created in advanced particle accelerators but no longer exist under normal conditions, gave way to quarks – the particles responsible for most of the mass in matter today. By the time it was one millionth of a second old, the Universe no longer had enough energy to make quarks. The production of lighter, more commonplace particles, such as electrons, took over for another fraction of a second before this, too, came to a halt. In less than a second, the Big Bang and inflation had created all the matter in the Universe today – approximately 10^{50} tonnes.

Ongoing process The tracks of subatomic particles form an intricate pattern in a bubble chamber filled with liquid hydrogen as particles and antiparticles spiral in opposite directions. The reactions brought about in modern particle accelerators mimic processes that occurred in the first split second of the Universe.

The elemental furnace

By the end of the first second of the Universe's existence, all the matter that exists today had been created. In fact, the Universe probably contained much more material than it does now. From the end of inflation for the first three minutes of its existence, these particles of matter formed a thick particle soup that gradually became more ordered and complex as the temperature dropped.

MATTER AND ANTIMATTER

The post-inflation Universe contained matter and antimatter in almost equal measures, and was so dense that particles and antiparticles were constantly colliding and annihilating each other. Today, however, the Universe is made entirely of normal matter. Antiparticles occasionally wink into existence around black holes or through radioactive decay, but they are destroyed on contact with normal particles, in the process releasing very distinctive gamma rays that betray their brief existence.

Where did all the antimatter go? The answer is that it was simply annihilated by coming into contact with particles of normal matter, releasing a storm of high-energy gamma rays that helped to keep the expanding Universe searingly hot. If matter and antimatter had been evenly balanced, the Universe would swiftly have become empty but, fortunately, there seems to have been a slight imbalance. Physicists have traced this back to the decay of a heavy, high-energy particle called the X boson. Both the X and its antimatter twin, the anti-X, tend to produce slightly more matter than antimatter when they decay. The result was that, for every 100 million particles, there were just 99,999,999 antiparticles. The modern Universe is made up of the tiny fraction of matter particles that survived annihilation.

PARTICLE SOUP

Since more massive particles require more energy to make them, the steadily dropping temperature of the Universe during its first second meant that the production of heavy, unstable particles soon stopped. Lighter particles (quarks and leptons) now began to dominate the Universe.

Quarks are the heavier of these. There are six types in all but only the lightest two, the up-quark and the down-quark, can survive in today's low-energy Universe. Up to 1 microsecond (one millionth of a second), the temperature of the Universe was still hot enough to produce quark-antiquark pairs spontaneously. Most were annihilated instantly, but quarks were also produced by decaying X bosons, so an excess of normal quarks survived.

Leptons are much lighter and smaller particles than quarks. Of the six types of lepton, three have significant masses, and only the lightest of these – the electron – is still found in matter today. The other three leptons are particles with very low mass called neutrinos. They are created in nuclear reactions and are virtually undetectable because they pass through most materials. Neutrinos generated in the Big Bang and more recently may account for a large part of the Universe's missing mass.

By the time the Universe was one second old, neutrinos overwhelmingly outnumbered all the other types of matter. Since they are lighter, leptons – and particularly the near-massless neutrinos – were produced for the longest period. But even these were swamped by the huge number of photons – particles without any mass at all that transmit light and other types of radiation. For every particle of matter in the primordial Universe, there were 100 million photons.

High-energy particles
Physicists study the conditions of the Big Bang using particle accelerators. Particles that cannot normally exist in today's low-energy Universe appear briefly as subatomic particles accelerated close to the speed of light, collide and release enormous amounts of energy. The particles leave tracks through a detector chamber, where their paths reveal properties such as mass and electric charge.

BLACK HOLES AND WIMPs

Mixed in with these familiar particles were more exotic species, which may have survived until the present. These include tiny black holes weighing a few hundred thousand tonnes, magnetic monopoles – heavy particles with a single magnetic pole instead of a north-south pair – and Weakly Interactive Massive Particles (WIMPs) – heavy but hard-to-detect particles that could also contribute to the missing mass. There may also have been cosmic strings: long, incredibly thin strands containing enormous amounts of energy and mass.

FIRST ATOMS

At around 0.1 milliseconds, the next stage of creation began. Temperatures had now dropped low enough for quarks to begin to stick together, forming groups of three particles held together by the strong nuclear force. Where two up-quarks and a down-quark met, they formed a positively charged proton. Where two down-quarks combined with a single up-quark, they formed an electrically neutral neutron. These two types of particle are the building blocks for the central nuclei of atoms. By the time the Universe was 1 second old, its temperature had dropped to around 10 billion °C (18 billion °F) and all the surviving quarks had been absorbed into protons and neutrons.

The next three minutes saw a period of frenetic activity which produced the first elements. Although equal numbers of protons and neutrons had been produced by the combination of quarks, neutrons were unstable, spontaneously transforming into protons until there were eventually seven protons to every neutron. As the temperature dropped below 1 billion °C (1.8 billion °F), it finally became cool enough for atomic nuclei to be stable.

Nuclear fusion now spread through the Universe, as colliding protons and neutrons stuck together to form the nuclei of helium, the light metal lithium, and heavy forms of hydrogen. For a few minutes, until the protons and neutrons became too widely scattered, the entire Universe was forging elements like the core of an immense star, several million kilometres across.

This model of nucleosynthesis (the manufacture of atomic nuclei) makes important predictions about the proportions of elements created in the early cosmos. Cosmologists can test these predictions by looking at the oldest surviving stars in globular clusters and elliptical galaxies. According to the theory, any surplus protons left over after nucleosynthesis would have become the nuclei of the simplest element, hydrogen. Nearly all the neutrons would have combined with protons to make helium and heavy forms (isotopes) of hydrogen called deuterium and tritium. And only a very few of these initial particles would have formed nuclei of lithium. Astronomers have found that 10 per cent of the atoms in the oldest known stars are helium, making up 23 per cent of their masses, as helium is much heavier than hydrogen. These are the exact proportions predicted by the model, and this discovery provides an important piece of evidence in favour of the Big Bang.

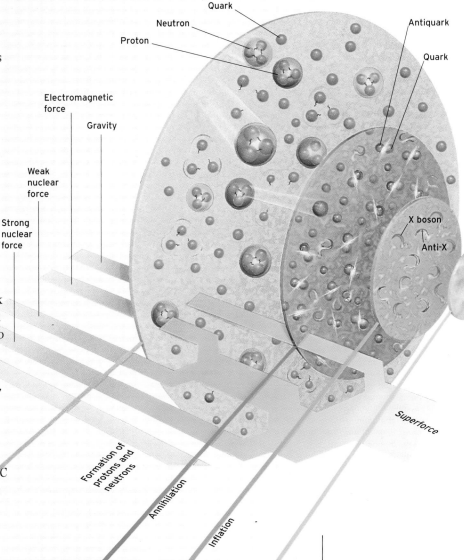

In the blink of an eye
The Universe's first second saw the Big Bang, inflation, the separation of forces, the creation and annihilation of matter and antimatter, and the fusion of quarks to form protons and neutrons.

The beginnings of structure

After the smoke had cleared... This artist's impression shows the period when the first stars, clusters, and galaxies began to condense from tiny irregularities in the fog of the early Universe.

At the end of its first three minutes of existence, the Universe was composed mainly of atomic nuclei, free-floating leptons and a blizzard of light-carrying photons. As it continued to expand and cool, the seeds of galaxy clusters and superclusters appeared.

After the fury of the Big Bang, inflation and nucleosynthesis, the Universe now settled down into a relatively calm period of expansion and cooling. This period is known as the radiation era because, at the time, most of the energy in the Universe was locked up in photons of radiation rather than matter. This phase lasted for about 300,000 years.

A DENSE BALL OF LIGHT

The early Universe was opaque – the matter within it was so densely packed that photons could only travel for tiny distances before colliding with an atomic nucleus or a free-floating electron and ricocheting off in another direction. As light could not travel over long distances, it would be impossible to see anything. If we could have observed the Universe from outside at this time, it would have appeared as an expanding incandescent ball – a giant sun growing at the speed of light.

Most of the radiation in the early Universe was in the form of gamma rays released by the mutual annihilation of matter and antimatter. But constant collisions with matter gradually transferred energy from the photons into the protons and neutrons – atomic nuclei – and electrons. As the energy of the photons dropped, they became X rays, then ultraviolet light and, finally, visible light and infrared heat radiation. At the same time, the energy transferred into matter kept the particles moving around rapidly, and stopped nuclei and electrons combining to form atoms.

MATTER AND LIGHT SEPARATE

After 300,000 years of steady expansion, the temperature dropped below an important threshold. At 3,000 °C (5,400 °F), electrons were finally able to combine with atomic nuclei. The Universe cleared and became transparent, as light was able to travel in straight lines. This event is called decoupling because, from this point on, matter and light were set free of each other. The Universe now entered the era of matter: 75 per cent of it hydrogen and most of the rest helium. Most of its energy since then has been locked up in the mass of its particles.

Decoupling is also important because it is the first event in the history of the Universe that we can actually see. Prior to this, the cosmos was opaque and we have no way of parting the veil to look back into the radiation era. But it should be possible to look back and trace the afterglow of this first visible event at the edge of the observable Universe. In fact, the echo of decoupling was found in the early 1960s by

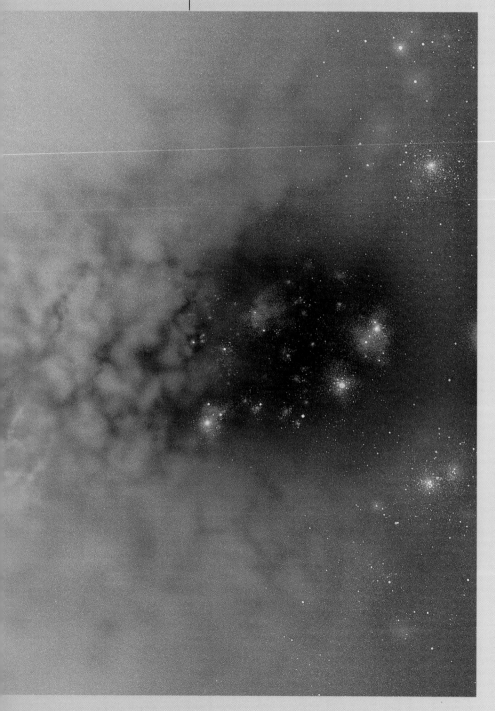

two astronomers who weren't even looking for it. Arno Penzias and Robert Wilson of Bell Telephone Laboratories were using sophisticated microwave antennae to scan the sky, looking for microwave radio signals from the edges of our galaxy. However, they found their measurements constantly beset by a faint microwave glow, apparently coming from all over the sky. They attempted to trace the source to everything from nearby radio transmitters to pigeon droppings, but without success. It was only when they discussed the problem with cosmologists that they realized what they had found – the radiation from the Big Bang, cooled and redshifted so that it glowed at just 3 °C (5.4 °F) above absolute zero.

THE GLOWING LIMITS OF THE COSMOS

This cosmic background radiation is probably the most important single piece of evidence for the Big Bang. Other theories of cosmology can deny or explain away redshifts and the expansion of the Universe, but none can explain convincingly why the limits of the observable Universe should glow in this way.

Since its discovery, the background radiation has been studied intensely. From the ground, radio astronomers found that it is remarkably even across the entire sky. This suggests that the Universe at decoupling was extremely uniform – just what the Big Bang theory requires to explain the large-scale uniformity of the cosmos today: the fact that equal numbers of galaxies are seen in every direction. However, this very smoothness troubled some cosmologists. There were none of the tiny irregularities that would have allowed matter to clump together and form galaxies. If the Universe had a completely smooth background radiation, we shouldn't be here to see it.

The mystery was solved in 1992, when a dedicated satellite called the Cosmic Background Explorer (COBE) completed a map of the whole sky's microwave background. COBE detected ripples of just a few millionths of a degree above or below the average temperature. The warm patches show areas where hot gas was concentrating, while the cold ones show comparatively empty regions. The pattern of ripples discovered by COBE is strongly reminiscent of the galaxy filaments and intervening voids found in the modern Universe.

THE SEEDS OF GALAXIES IN DARK MATTER

The cause of these all-important ripples is hotly debated, but many astronomers believe that the answer is to be found in the Universe's enigmatic dark matter. Up to the point of decoupling, constant collisions with photons of light would keep particles of normal matter moving around and prevent them from clumping together, so the ripples cannot be formed from matter. But dark matter, such as WIMPs and black holes, would have been affected only by gravity. If there were tiny fluctuations in the distribution of this matter created in the Big Bang itself, then by the time of decoupling they could have begun to form more dense or less dense regions, which would later have pulled normal matter in around them.

However, cosmologists cannot agree on how the smooth Universe created by the Big Bang developed fluctuations in the first place. This is, again, a subject of fierce debate, but the most widely accepted theory is that before inflation, the tiny Universe was riddled with minuscule variations caused by quantum effects. Inflation made these fluctuations in density trillions of times bigger, turning them into the seeds of galaxy clusters and superclusters.

COBE all-sky map
In COBE's map of the background radiation, warm, dense areas are shown in pink and cool, emptier regions in blue. The maximum temperature difference between bright pink and dark blue areas is just 0.0005 °C (0.0003 °F).

In the aftermath of decoupling, the Universe was plunged into a dark age from which it did not emerge for millions of years. In the darkness, matter began to condense, ultimately forming the first galaxies. But astronomers still disagree about how the creation of the galaxies began.

Galaxy formation is one of the most exciting areas of cosmological research. This era lies just beyond the reach of modern telescopes. The Hubble Space Telescope can look back through space and time to an era one billion years after the Big Bang where it has seen irregular galaxies apparently merging together to form spiral galaxies, but by this time the Universe was already fairly old. The answers to many fundamental questions about galaxy origins may have to wait until NASA's Next Generation Space Telescope, the successor to the HST, is launched.

HOW GALAXIES FORM

There are two main theories of galaxy formation – top-down and bottom-up. In the top-down theory, gas clouds collapse and begin to rotate, forming protogalaxies. Individual stars begin to take shape, and material sinks toward the centre of the galaxy, where it may form a supermassive black hole as heavy as a billion Suns. In the bottom-up model, the black hole forms first and pulls in gas from around it. This last theory has gained ground recently. It now seems that all large galaxies contain giant black holes at their centre, indicating that they could be key to a galaxy's evolution. In addition, distant quasars far back in time seem to be powered by black holes that are pulling in material from the young galaxies around them.

These black holes may have formed as an inevitable result of inflation. Alternatively, they could simply have formed from collapsing gas clouds. A slow and steady collapse can form a black hole just as effectively as the violent implosion of a supernova's core – the only requirement is that a region of space becomes so dense that light cannot escape its gravity.

If galaxies formed around black holes, they would have emitted fierce X-rays as material poured into the hole and was heated and ripped apart by tidal forces. This radiation might be the solution to another cosmological mystery. At some point after decoupling, when all the matter in the Universe had joined together into atoms, something ionized the gas atoms in intergalactic clouds, splitting them apart again. The only explanation is intense radiation from some source or other, and the searing light from early quasars is currently the prime suspect.

VIOLENT ORIGINS

Although quasars are among the most energetic objects that we can currently see, their violence is dwarfed by other theoretical objects that may have existed in the early Universe. Possibly the most spectacular of all these theoretical objects, are the megasuns – vast stars with 200 to 300 times the mass of the Sun, which could only form in the special conditions that existed in the Universe's youth.

Hubble Deep Field
This Hubble Space Telescope image looks back 11 billion years, into the early Universe, when irregular blue galaxies were merging to form larger, more complex systems. These galactic building blocks probably formed by coalescing around central black holes.

Today, the largest stars – so-called Wolf-Rayet stars – may grow up to about 100 solar masses. But as they accumulate more mass, they become increasingly unstable. The radiation from these heavyweights creates such strong stellar winds that they blow away up to half of their mass during their short lifespan. Above a certain limit the star's radiation is so powerful that it drives off any nearby material that might otherwise have increased its mass. Furthermore, these giant stars are rich in heavy elements such as carbon, which increase their rate of burning and help to make them unstable. In the early Universe, however, there were no heavy elements in existence, so the stars that formed

No one has ever seen the death throes of such a massive star, but one of two things might happen. If the star is below 270 solar masses, it goes hypernova – when the core collapses it becomes so incredibly hot that it triggers a vast nuclear firestorm, burning through all the star's available material in an outburst a hundred times brighter than a modern supernova. The outward blast of radiation is so savage that not even a black hole survives. Hypernovae are another possible explanation for the ionized gas in between the galaxies. They may also explain why many of these distant clouds, which should not have changed their composition since the Big Bang, contain small traces of heavy

were all made of pure hydrogen and helium. This meant that they used their fuel at a more sedate rate—although they would still have been 10 million times more luminous than our own Sun – and could have grown considerably larger than today's most massive stars.

LIFE AND DEATH OF A MEGASUN
Some astronomers believe that an early generation of megasuns were the first objects to light up the Universe as it emerged from its dark age, before galaxies had even begun to form. These stars would have raced through their life cycles in just a few million years, converting hydrogen and helium in their cores into heavy elements on a massive scale. Unlike today's violent Wolf-Rayet stars, their stellar winds would not have been strong enough to blow away much material. They would have ended their lives with nearly the same mass with which they began them.

elements. Shockwaves from hypernovae could even have triggered the formation of galaxies.

The death of a megasun weighing more than 270 Suns is even more cataclysmic. Above this mass, the core's gravity overcomes the pressure of radiation and, when its nuclear reactions cut out, it collapses into a 30-solar-mass black hole that consumes the star from the inside. The star's material spirals inward, forming a hot accretion disc similar to those found in quasars. But the black hole is swallowing far more material in a much shorter time than a normal quasar and is consequently much brighter. A collapsing megasun could shine for a few seconds with the light of 10 billion galaxies— as bright as the Universe itself. However, since most of its energy would be released in the form of gamma rays and then redshifted into X-rays, these theoretical objects have not yet been seen. If they are found, we will be one step closer to understanding the origins of the Universe.

Stellar colossus This artist's impression shows a brilliant, short-lived megasun, surrounded by the nebulosity from which it condensed. The star will shine for only a few million years before self-destructing in a gigantic explosion.

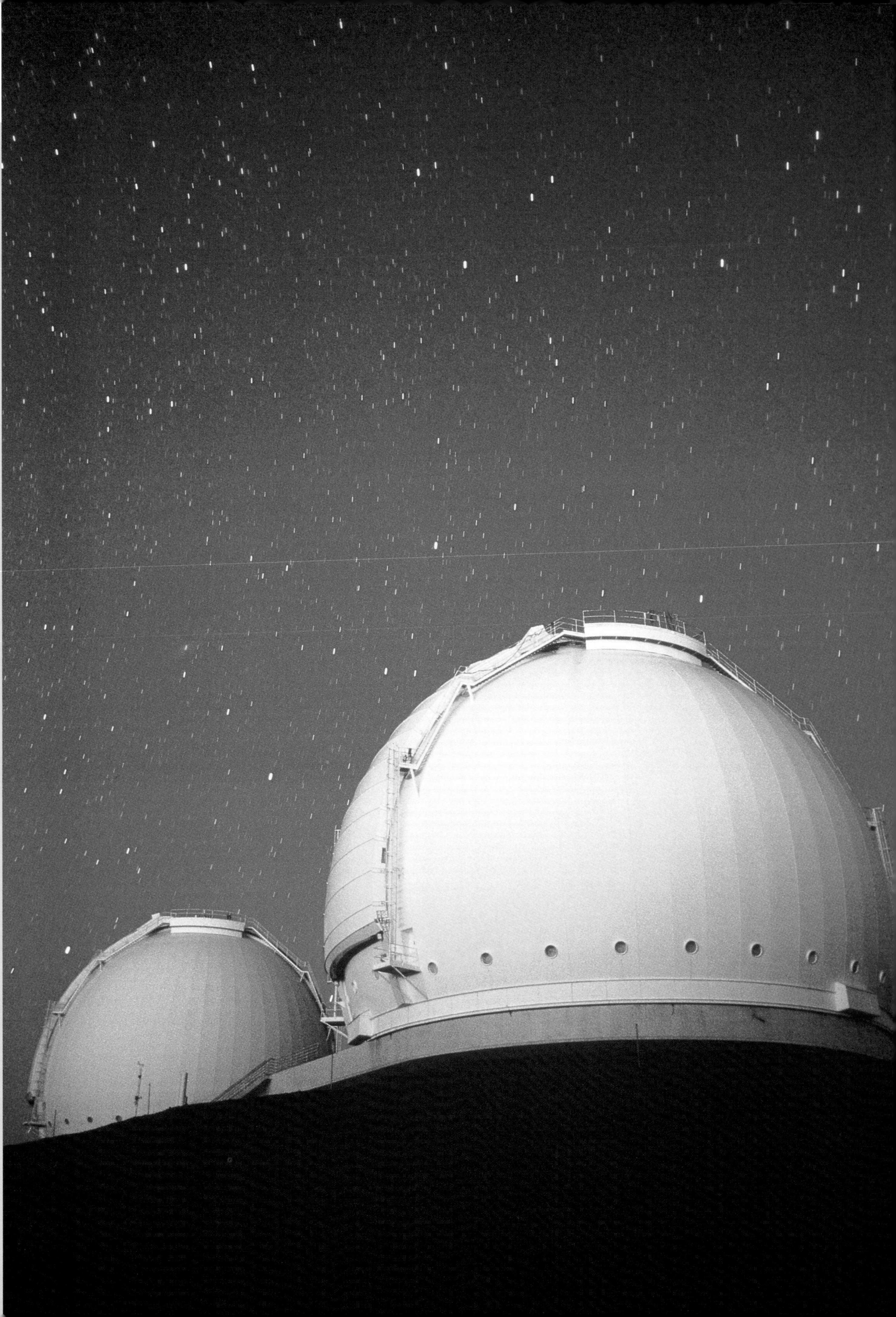

How astronomers work

Modern astronomers have come a long way from the astronomer-priests of the ancient world, and even the 19th-century enthusiasts who relied on personal wealth or generous patrons to pursue their work. Today, most work in academic institutions and are largely funded by government grants.

Astronomy today uses a wide variety of different techniques to probe the secrets of the Universe. Professional astronomers are as likely to be found scribbling cosmological calculations on a blackboard, or running complex computer simulations of colliding galaxies and evolving stars, as they are to be anywhere near an observatory. Even when they are collecting data from the sky, electronic instruments have long since replaced human observers at the business end of the telescope.

Astronomy is a science like any other, and as such has to obey certain rules. All sciences involve the analysis of data from experiments – be it a chemical reaction in a laboratory or a series of measurements taken from distant stars – and the formation of theories to account for the results. For a discipline to be considered a true science, its experiments should be objective and reproducible by anyone who follows the same method, and its theories should be concise and testable. The theories should also generate predictions that can be put to the test– in astronomy's case by further observations.

THEORY REINFORCED BY OBSERVATION

Any theory put forward in astronomy is just that – a theory. It holds sway only as long as it can explain any new data that comes to light. For example, traditional Newtonian gravity could not explain perturbations in Mercury's orbit, but Einstein's theory of general relativity could. Relativity also predicted the phenomenon of gravitational lensing – the bending of light as it passes near massive objects – and observations later confirmed it. Relativity therefore replaced Newton's old theory – although Newtonian mechanics still works perfectly well for everyday situations. Einstein's ideas have survived for nearly a century now, and so far nothing has appeared to prove them wrong. Today, we can say with some confidence that Einstein, like Newton, is unlikely to be proven completely wrong, although general relativity may prove to be an oversimplification of some yet-deeper fundamental laws.

On a smaller scale, astronomers are testing and overthrowing theories in this way all the time. Although they are better equipped than ever before, resources are still limited, so they tend to direct their research toward answering specific questions. To do this, they devise experiments just like any other scientist. These may range from a programme of observations to gather data with one of the world's major telescopes to the launch of an orbiting observatory or spaceprobe to another world.

Astronomers typically spend just a few days or nights gathering information, and many months analysing it and drawing conclusions. The end result is a scientific paper – a formal publication in a scientific journal outlining the experiment, its results, and any conclusions that may be reached. Before publication, papers are circulated to other astronomers in the same field for comments and criticism. This review process is supposed to ensure that results and theories reaching publication meet certain standards. This means that the results can be accepted by the astronomical community without the need for testing and replication, although the conclusions drawn from those results may be the subject of contention.

The Keck telescopes These 10-m (33-ft) twins (far left), sited in Hawaii, are among the most powerful telescopes in the world.

Searching the skies An astronomer at work in the control room of the 100-m (330-ft) Max Planck Institute radio telescope near Bonn, Germany.

Hubble Space Telescope The HST (centre) is a medium-large-sized telescope situated above the atmosphere's distortions. It has a clear view of the Universe unmatched by any other instrument.

Observatories and telescopes

Dating back to the small, primitive devices used by Galileo and his contemporaries, the telescope is astronomy's single greatest tool. Modern astronomical telescopes are computer-controlled leviathans capable of gathering light from the far reaches of the Universe and analysing it in many different ways.

BIG, BIGGER, BIGGEST

Professional astronomers today use reflecting telescopes instead of lens-based refractors. A reflector has a large concave primary mirror to collect light and reflect it to a smaller, secondary mirror. From here, light may be reflected several times more before it reaches a focus. It can be detected by any number of different instruments, or even split into two or more beams and passed to several instruments simultaneously.

The larger a telescope is, the greater its light grasp – the amount of light it can collect – and the brighter the resulting image. Larger telescopes also have higher resolution – they can detect finer details. Resolution is governed by the distance between opposite sides of the light-collecting surface. In theory, large telescopes can resolve well below a second of arc ($^1/_{3,600}$ degree), while the human eye can only see details a few minutes of arc ($^1/_{60}$ths of a degree) across. However, disturbances in the atmosphere always blur a ground-based telescope's view and prevent it from operating at its theoretical resolution.

Until the 1990s, there seemed to be an upper limit on the size of a reflecting telescope. The primary mirror has to be made with great precision to reflect all the light it collects to the same point – variations the thickness of a human hair could create a blurred image – and even in small telescopes the mirror's own weight can cause it to distort. Passive frameworks were designed to support the mirror and keep it in shape, but even these would not work above a certain size. For decades, the largest, successful single-mirror telescope was the 5-m (200-inch) reflector on Palomar Mountain in California.

However, a new generation of giant telescopes found an ingenious solution – computer-controlled support systems that constantly adjust, flexing the mirror into its ideal shape. This concept, called active optics, is used to support both large single mirrors and rings of multiple hexagonal mirrors – slotted together in a honeycomb pattern to form a perfect overall shape. Active optics has given rise to a new generation of giant telescopes, including the Very Large Telescope in Chile and the Keck I and Keck II instruments in Hawaii.

Although larger mirrors increase a telescope's light grasp, turbulence in Earth's atmosphere can still distort the images of distant stars. Astronomers call the quality of the atmosphere "seeing." By siting telescopes on high mountains above the thick lower atmosphere, observatories can achieve the best possible seeing, and also observe wavelengths that are absorbed by the time they reach sea level.

New methods can improve resolution even further. Laser beams can project an artificial reference star into the upper atmosphere close to the area being studied. By monitoring the changes to this predictable light source, a computer can adjust the shape of the telescope's main mirror to compensate for any atmospheric distortion.

Another method of improving resolution is to link two or more telescopes together by a complex electronic technique called interferometry. The instruments behave as if they were positioned on the edges of a single vast telescope. Interferometry is used to combine Hawaii's twin Keck telescopes to form an instrument

Dawn at the Very Large Telescope The VLT, sited on a Chilean mountaintop, combines four giant reflectors with several smaller telescopes to create an instrument with the largest light grasp and finest resolution available in the world today.

resolution of a CCD camera depends on the number of pixels it contains, and such devices have only come close to the resolution of traditional photographic film in the 1990s.

But CCDs have many other advantages. One is their usefulness in photometry – the precise measurement of the amount of light coming from an object. Unlike a photographic plate, a CCD can be read and reset as often as required. The charge on the CCD each time will be in exact proportion to the amount of light coming from a star, so a computer can plot a light curve of the changing amount of light the CCD detects. Photometry is particularly useful for studying variable stars, and the atmospheres of distant planets – recording changes to the light of a distant star as it is eclipsed or occulted by the planet. It can also reveal the shapes of asteroids, since an odd shaped asteroid will reflect varying amounts of sunlight as it rotates.

A third popular method of analysing light from a telescope is spectroscopy – splitting light into a spectrum and recording the dark or bright lines produced in light from a star or a nebula. These lines, caused by atoms absorbing or emitting light with specific colours, can reveal the chemical content of stellar atmospheres or interstellar gas clouds.

Spectrographs are instruments attached to the end of a spectroscope to record the spectra it produces. As with normal cameras, spectrographs once relied on photographic film but have now been largely replaced by CCDs. Instead of using glass prisms, which would absorb much of the faint starlight, astronomical spectroscopes split the light apart using a diffraction grating (a thin glass plate etched with thousands of very fine lines). As the light passes through the clear slits in the plate, it spreads out in ripples. Diffraction gratings split the different colours of light much more widely than traditional prisms, making the separation of neighbouring spectral lines much simpler.

Liquid mirror telescopes This revolutionary NASA telescope uses a 3-m (10-ft) mirror made by a slowly spinning dish of mercury.

with 85 m (280 ft) of resolving power. Greater still is the Very Large Telescope in Chile, with four 8.2-m (27-ft) telescopes and three 1.8-m (6-ft) instruments, and the combined resolution of a telescope more than 120 m (390 ft) across.

CAPTURING LIGHT FROM SPACE

Once a telescope has collected light, it can be detected and analysed in several different ways. The most popular methods are imaging, photometry and spectroscopy.

Often the observer's place at the telescope's eyepiece is occupied by a camera, which projects the image formed by the telescope onto a light-sensitive surface in order to record it. Rather than using traditional photographic film, the charge-coupled device (CCD), similar to the technology found in a digital camera or webcam, is often used. A CCD is a special type of electronic detector that converts the light rays striking its surface into an electrical charge stored in a grid of picture elements or pixels. The amount of charge that builds up on each pixel over the length of an exposure can be read by a computer and built up into an image. The

Massive mirrors The largest individual telescopes in the world are the twin Keck Telescopes, at Mauna Kea, Hawaii. Each of these giants has a collecting area equivalent to a single 10-m (33-ft) diameter mirror. Other even larger telescopes are now being planned.

The invisible Universe

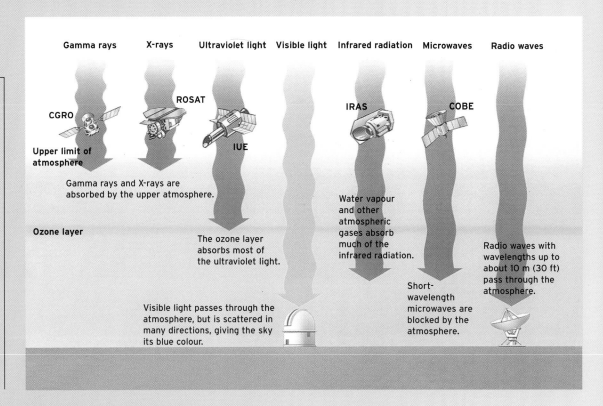

Gamma rays   X-rays   Ultraviolet light   Visible light   Infrared radiation   Microwaves   Radio waves

CGRO

ROSAT

IRAS

COBE

IUE

Upper limit of atmosphere

Gamma rays and X-rays are absorbed by the upper atmosphere.

Ozone layer

The ozone layer absorbs most of the ultraviolet light.

Water vapour and other atmospheric gases absorb much of the infrared radiation.

Radio waves with wavelengths up to about 10 m (30 ft) pass through the atmosphere.

Visible light passes through the atmosphere, but is scattered in many directions, giving the sky its blue colour.

Short-wavelength microwaves are blocked by the atmosphere.

A protective shroud

Earth's atmosphere blankets the globe, protecting us from many of the harmful forms of electro-magnetic radiation from space. Unfortunately for astronomers, this prevents them from collecting many of the emissions from celestial objects. To detect these, astronomers place satellite observatories in orbit high above the atmosphere.

Visible light is just a small section of the electromagnetic spectrum. Using specialized instruments, astronomers today can study the Universe using many other forms of this radiation, from high-energy gamma rays to low-powered radio waves.

An electromagnetic wave is defined by its wavelength, the distance between the peaks or troughs, and its frequency, the number of waves travelling past any point in each second. The wave's speed – the distance travelled by any point on the wave in one second – is its wavelength multiplied by its frequency.

But because all electromagnetic radiation travels at the fixed speed of light, as the frequency of a wave increases, its wavelength must decrease. Energy increases as frequency increases, so high-temperature objects with high energies produce short-wavelength radiation. Conversely, cool objects with little energy emit long-wavelength, low-frequency radiation.

RADIO ASTRONOMY

Radio waves have the longest wavelengths – ranging from kilometres down to 1 mm ($\frac{1}{25}$ in). Radio astronomers were the first observers to map the sky at non-optical wavelengths. Radio waves let astronomers study some of the coolest, lowest-energy objects in the galaxy – such as the cosmic background radiation – the afterglow of the Big Bang itself, at just 3 °C (5 °F) above absolute zero.

The longer the wavelength of electromagnetic waves, the lower the resolution any particular telescope can achieve. To gain resolution, and to increase sensitivity, radio astronomers build large telescopes and link widely spaced instruments into arrays. These form some of the biggest scientific instruments on Earth.

In a radio telescope, a huge metal dish gathers and reflects radio waves, focusing them onto a feed horn. There, the waves are converted into electrical signals that astronomers can analyse in detail. Some radio telescopes – often the largest ones – are able to view only part of the heavens, while others can point freely anywhere in the sky.

However, even the largest radio telescope in the world – the 300 m (1,000 ft) dish at Arecibo, Puerto Rico – has very low resolution compared with optical telescopes. So radio astronomers developed interferometry – electronically combining the signals from two or more telescopes. This massively increased the resolution, though not the collecting area, of radio telescopes. Arrays of telescopes linked by interferometry now stretch around the world and will shortly be extended into space.

INFRARED TELESCOPES

The infrared is the region of the spectrum between the wavelengths of the reddest visible light (about 700 nanometres—nm—or 700 billionths of a metre) and the shortest radio waves (about 1 mm). It is often called heat radiation, since it is emitted by objects too cool to glow at visible wavelengths. Infrared radiation can help reveal warm regions hidden behind dark dust clouds, such as the centre of our galaxy, map the hearts of nebulae and find the locations of faint brown dwarf stars.

Collecting infrared wavelengths presents a problem – they are absorbed by water vapour in the lower atmosphere and swamped by the instruments themselves. Infrared telescopes are therefore sited on high mountains, or in space, and use detectors chilled with liquid helium.

ULTRAVIOLET AND BEYOND

Beyond the blue end of the visible spectrum lies ultraviolet light with wavelengths from 390 to 10 nm, emitted by extremely hot objects. Earth's atmosphere actually blocks out these wavelengths so they can be observed only with space-based telescopes. All stars emit some ultraviolet, and some hot, young and massive stars emit most of their energy in this region.

Beyond ultraviolet light lie X-rays, with wavelengths less than 10 nm, and gamma rays, with wavelengths below 0.01 nm. X-rays and gamma rays pass straight through mirrors – the only way to focus X-rays is by using a metal reflector that the rays strike at a shallow angle and ricochet off onto an electronic detector. Gamma rays cannot be reflected at all, so the only way to map their sources is to use a heavily shielded detector, which rays can only enter if they are travelling from a particular direction.

Both these types of radiation come from the hottest and most violent regions of space – the Sun's upper corona, supernova remnants, the hot gas in galaxy clusters, and material spiralling into black holes.

Very Large Array (VLA) This radio telescope array in New Mexico consists of 27 individual 25-m (81-ft) dishes, mounted on a Y-shaped track system that allows them to function as a single telescope with a diameter of up to 36 km (22 miles).

Neutrino observatories
Some astronomers study particles instead of radiation. At Sudbury Neutrino Observatory in Ontario, Canada, an array of detectors 2,100 m (6,800 ft) below ground monitors the arrival of high-speed neutrinos from sources such as supernovae.

Amateur astronomy

While the fascination of exploring the wonders of the Universe on a clear night is shared by all amateur astronomers, some people catch the astronomy bug badly, and want to take things farther. Fortunately, astronomy is one of the few remaining sciences where amateurs, sometimes equipped with little more than a pair of binoculars and a notebook, can still make an important contribution.

Although there are more professional astronomers working today than at any previous time in history, most are so busy writing scientific papers or applying for telescope time that they rarely have time to look at the night sky. Furthermore, most of them are highly trained specialists who might know everything there is to know about Cepheid variables, but very little about how to recognize a new comet.

Be prepared This observer's tool kit consists of a clipboard with a red-filtered light clamped on, pens and pencils attached by string, and a sturdy clip to hold paper in place. Predesigned reporting forms for observations are issued by many astronomical societies.

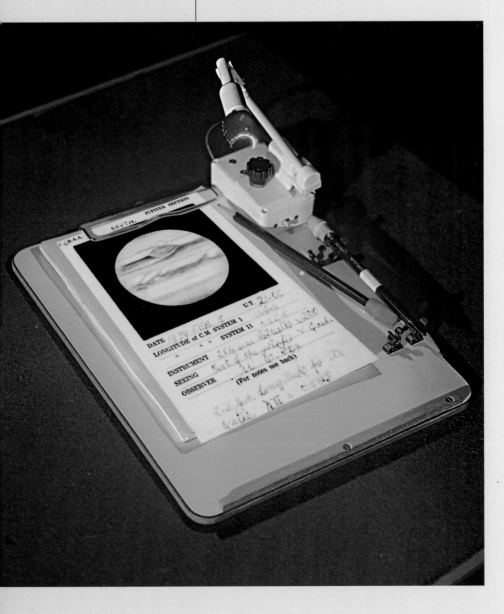

WHAT THE AMATEUR CAN CONTRIBUTE

Amateur astronomers can function as extra pairs of eyes for the astronomical community, keeping regular watch for new phenomena and alerting the world so the major observatories can swing their telescopes into action. Since Solar System objects are nearby, bright and relatively easy targets for a small telescope, many amateurs choose to concentrate on them. Although they offer less scope for scientific discovery in these days of orbiting telescopes and spaceprobes, they can still provide beautiful and fascinating views and, on occasion, a few surprises. In some cases, a new discovery can bring fame and celebrity to a lucky amateur.

Comet hunters spend every available night sweeping the sky for new comets on their approach to the Sun. In order to avoid false alarms, they must familiarize themselves with a host of misleading star clusters, nebulae and galaxies. On the rare occasions when someone spots a new comet, as with any other new astronomical discovery, they must notify the Central Bureau of Astronomical Telegrams. If the discovery is confirmed and does not match the orbit of any previously known object, then the new comet is named for its discoverer.

In its early days, NASA relied mostly on maps of the Moon compiled by dedicated amateurs. Occasional glimpses of strange orange glows called transient lunar phenomena (TLPs), usually reported by amateurs, may yet reveal that the Moon is more active than was once thought. Most of the major astronomical societies have sections dedicated to collating observations of each major planet, and amateur astronomers are often the first to notice dust storms on Mars or changes in the clouds of Jupiter, for example.

STORMS AND SUPERNOVAE

Meteor watching can be a rewarding pursuit, and requires little more than a notebook and sky map. International meteor watches are often organized in advance of predictable storms such as the Leonids. When thousands of individual meteor counts are collated from around the world, astronomers can piece together the precise shape of a meteor stream.

Another popular pursuit for the more serious amateur is nova and supernova hunting. New novae are often very difficult to identify from other stars unless they are particularly bright or appear in the midst of very distinctive star patterns. As with the comet hunters, anyone seriously searching for novae requires a deep knowledge of star patterns. However, once again, the only real equipment a nova hunter needs is a good star atlas and a pair of binoculars.

Supernovae are a different matter, as they are extremely bright but also extremely rare. Any supernova within our own galaxy would appear truly brilliant, and be obvious to even a casual stargazer. Supernovae in other galaxies are very distant, but are still bright enough to alter a galaxy's appearance significantly to someone already familiar with it. Dedicated supernova hunters use powerful telescopes as well as detailed photographic atlases of the galaxies they are sweeping. Although they do not get the honour of naming a supernova, they can make a significant contribution to our knowledge of the Universe, because supernovae are valuable tools for establishing the distances of other galaxies.

WHAT YOU WILL NEED

Even if your aims are not scientific discovery but a wider appreciation of the cosmos, you may find some accessories useful – a comfortable chair, notepad and pencil to record and sketch your observations, and a thermos of something hot to drink. The average person takes about 20 minutes to adjust fully to a dark night, so to avoid ruining your night vision, carry a flashlight covered with red paper or plastic so you can see – our eyes are naturally much less sensitive to red wavelengths. Another useful accessory is a planisphere – a device consisting of two plastic discs which, when rotated, can show the night sky for your latitude at any time of year.

Once you have learned to find your way around the night sky, you might try astro-photography – taking pictures of the stars. At its simplest, all you need is a tripod-mounted camera capable of long, manually released exposures, a cable release to avoid touching the camera, and fast film. By pointing the camera at the sky and exposing the film for a few minutes, you should be able to make images of star trails as the sky slowly spins around the celestial poles – you may even capture a meteor or satellite trail.

IMAGES FROM YOUR TELESCOPE

You can attach your camera to a telescope with a simple device called a T-adaptor, and the results can be spectacular. However, because the telescope magnifies the movements of the sky, as well as the stars themselves, you will need a motor-driven telescope capable of steady motion to counter Earth's rotation. In addition, a working knowledge of specialty films, processing techniques, and so on, is needed to get the best results.

However, the recent appearance of high-resolution digital cameras has opened up the possibilities of astrophotography without a home darkroom or special films. Digital imaging makes astrophotography much simpler, though getting good results will require some skill and a lot of practice. Computer image-processing programs make it easy to fine-tune your images and combine photographs taken through different-coloured filters. Some amateurs have even found that webcams make ideal astrocameras.

The amateur telescope
This refracting (lens) telescope is fixed on an equatorial mount to keep track of the stars. A star diagonal at the observer's end diverts the light, allowing the eyepiece to be placed more conveniently, and a small finderscope of low-power magnification is attached to the side. The whole thing is mounted on a sturdy wooden tripod to prevent shaking.

Once familiar with the naked-eye stars, most people want to take a closer look with some kind of observing equipment. The options are bewilderingly varied, and you'll do best to buy from an established dealer.

BINOCULARS AND LENSES

Binoculars are often a far better purchase for the amateur on a budget than a really small telescope. They work by the same principle as refracting telescopes, collecting light and magnifying an image with a series of lenses.

Binoculars and refractors both have a lens at either end – a large objective, and a smaller eyepiece. The objective collects the light and brings it to a focus, forming an image which the eyepiece then magnifies. In addition to their lenses, binoculars contain glass prisms that fold the path of the light passing through them, and reduce the length of the tube.

Because the paths of light cross at the focus, they produce an upside-down image. This can be corrected with a third lens between objective and eyepiece, fitted as standard in binoculars and terrestrial telescopes, but usually left out of astronomical instruments where it would absorb precious starlight. Astronomers do not usually worry, since there is no real "up" in the sky, but a cheap attachment can be bought that corrects a telescope's image for other uses.

Binoculars are usually described by two numbers – the objective diameter and the magnification. For instance, 10x80 indicates 10x magnification and 80 mm diameter. As a rule, the diameter in millimetres should be at least 7 times the magnification. Ideal for viewing star clusters and other large objects, 10x80s provide a wide field of view, good light grasp, and low-enough magnification to avoid shaking when held in the hands.

AMATEUR TELESCOPES

Astronomical telescopes are of two basic types – lens-based refractors and reflectors. In a reflecting telescope, a concave primary mirror collects light and bounces it back to the eyepiece via a secondary mirror small enough to avoid obscuring a large area of the primary. Unlike refractors, reflectors come in a wide variety of different shapes and sizes.

When purchasing a telescope, there are several factors to consider. The most important are light grasp and resolution – any magnification quoted on the box or in literature should only be considered in relation to these, as high magnifications are useless if limited by blurry resolution. Increased light grasp gives a brighter image, while higher resolution creates a sharper one, and both these factors improve rapidly as the diameter of the objective lens or primary mirror increases.

Since mirrors are easier to make than lenses, a reflector of a given price will often have a diameter nearly twice that of a comparable refractor, making it a much better purchase for amateurs. In order to be useful, a lens telescope needs a diameter of at least 75 mm (3 inches), and most refractors on the market are far smaller than this. Steer clear of small, cheap instruments boasting high magnifications without providing other details – often they will be mass-produced with poor-quality optics and misleading packaging.

Reflectors, on the other hand, are usually made by dedicated companies producing high–quality goods. For the same price as a reasonable refractor, one can purchase a 150-mm (6-inch) reflector, with four times the light grasp and twice the resolution. Although the larger mirror makes the reflector bulkier, the length of the main tube can be shortened by reflecting the light off multiple mirrors. The highly popular Schmidt-Cassegrain and Maksutov designs bounce the light off the secondary mirror down through a hole at the centre of the primary and into the eyepiece.

EYEPIECES AND ACCESSORIES

The eyepiece is the lens – or series of lenses— responsible for magnification. Any instrument worth buying should offer a range of interchangeable eyepieces, which may be purchased separately. Good eyepieces generally

The refracting telescope This small refractor is equipped with an equatorial mount, with an SLR camera piggybacked onto it. A quality heavy-duty mount and tripod are as important as the telescope's optics. At medium power, check that all vibration dies out in about one second.

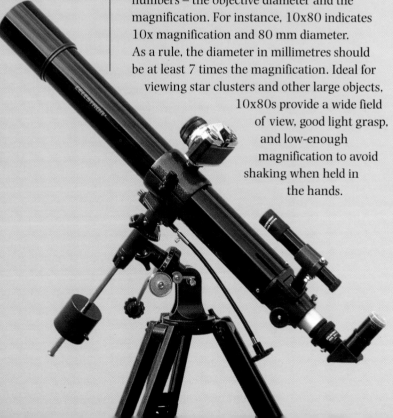

have barrels with diameters of at least 32 mm (1.25 inches) – avoid telescopes that have eyepieces with smaller barrel diameters.

Eyepieces magnify the image formed by a telescope's main optics. The power any eyepiece delivers is found by dividing its focal length (usually marked on the barrel) into the focal length of the telescope. For example, take an 200 mm $f/4$ reflector and a 30 mm eyepiece. The telescope has an 200 mm diameter main mirror and a focal length four times the mirror's diameter, or 800 mm. Dividing 800 by 30 gives a magnification of about 27x. With the same telescope, a 7.5 mm eyepiece yields a magnification of 108x.

Working out the magnification provided by any eyepiece therefore requires some basic maths, but bear in mind that the telescope's resolving power is ultimately governed by its diameter. As a rule of thumb, maximum usable magnification is usually about 35x per 10 mm of diameter.

Most telescope dealers will sell a range of eyepieces. They should also stock filters, which block certain colours and enhance others. Some filters enhance nebulae, some show features on other planets, and others simply block out the glow from streetlights. Another important accessory, if you want to use your telescope for astrophotography, is a T-adaptor, which allows a camera to screw onto the eyepiece.

TELESCOPE MOUNTINGS

The mount of a telescope – the mechanism that attaches it to a tripod or stand – is very important. A poor mount may wobble, and make the instrument unusable at high magnifications. The simplest type – often found on small refractors – is the alt-azimuth mount, which allows the telescope to swing horizontally and

vertically. This makes it easy to find objects, but tracking them requires adjustment of two separate axes.

Equatorial mounts avoid this problem by directing the axis of the mount toward the celestial pole. Up-down movement shifts the telescope through declination, and side-to-side movement shifts it in right ascension. Objects can then be located according to their celestial coordinates, and tracking requires just a simple movement through right ascension to keep up with the rotating sky. Equatorial mounts can even be motorized so they track the stars automatically.

In the past few years, a new type of mount has appeared – the computerized GO-TO mount. This cradlelike mechanism needs none of the careful alignment of an equatorial mount. The observer simply inputs their location and time, points the telescope at a handful of bright stars, and computer-driven motors do all the necessary tracking. In addition, the system carries a library of objects that it can point to at the touch of a button. All this technology makes GO-TO telescopes considerably more expensive, but many traditional mounts now also offer some degree of computerization.

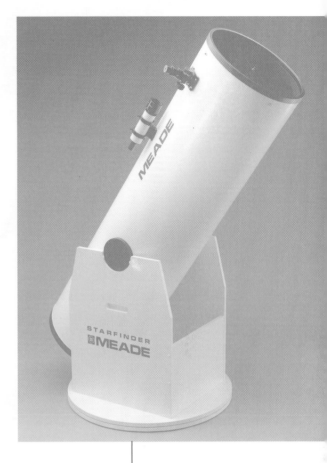

Dobsonian mounts Large amateur reflectors frequently use a simple cradlelike alt-azimuth mount called a Dobsonian. At low magnification, they have such wide fields of view that most objects can be found just by scanning across the sky.

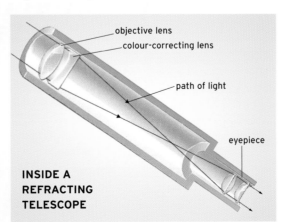

eyepiece

flat secondary mirror

path of light

concave primary mirror

INSIDE A NEWTONIAN REFLECTOR

objective lens

colour-correcting lens

path of light

eyepiece

INSIDE A REFRACTING TELESCOPE

Reflector versus refractor Excellent refractors and reflectors are available on today's market. But take the time to study the reviews and ads in astronomy magazines before buying – you'll be more informed and able to make a better choice.

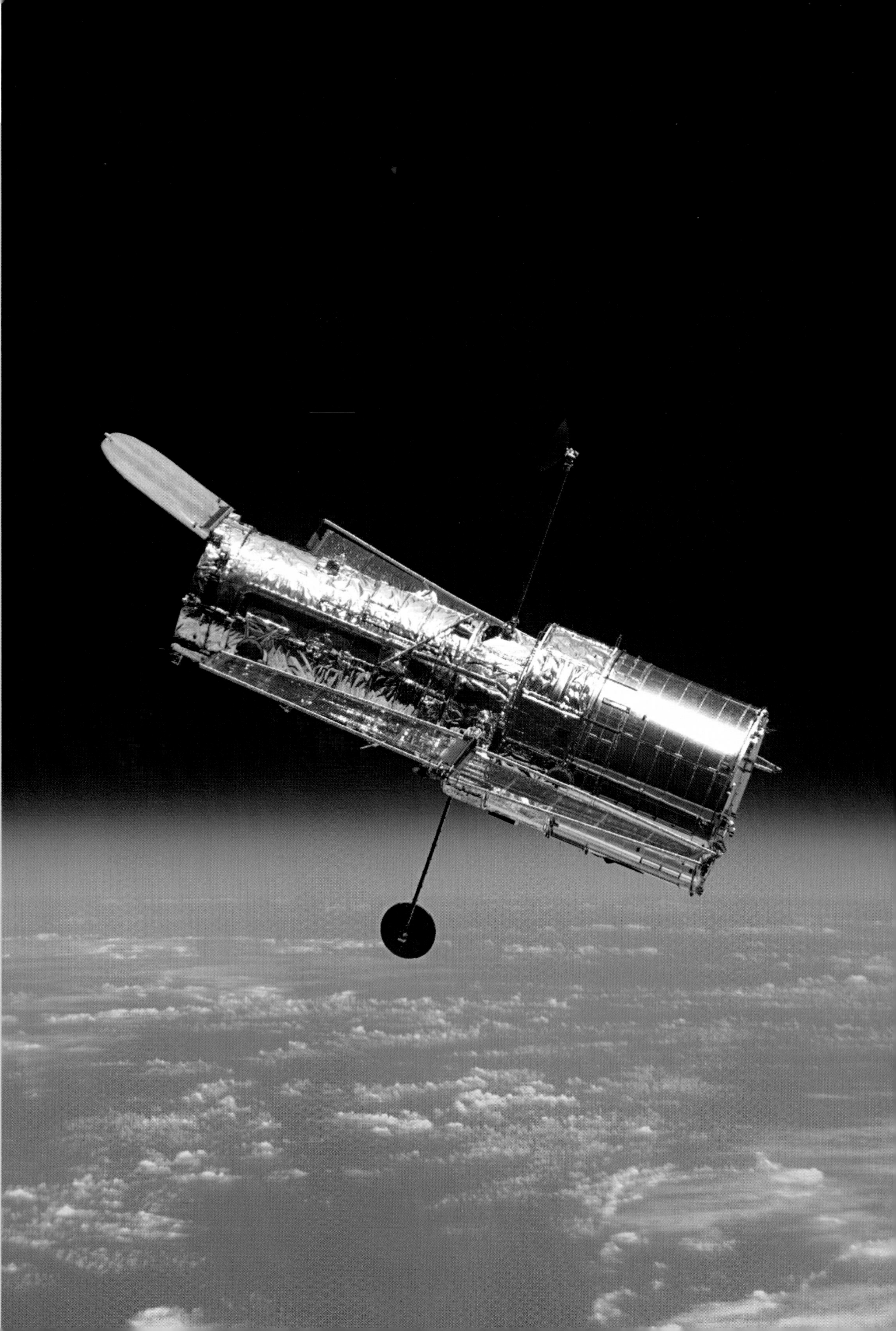

The history of astronomy

People have always been fascinated by the heavens, and interest in astronomy dates back to before written records. It is the oldest and most ambitious of all sciences and is still giving rise to astounding new discoveries, constantly reshaping the way we view the Universe.

Ancient people recognized that the cycle of the seasons affected their crops, and used the stars as timekeepers. It was not a great step from this to a belief that the celestial bodies had an influence over other aspects of their lives – and so astrology, the disreputable ancestor of astronomy, was born.

Astrology can be seen as the first scientific theory of the Universe. It was a genuine attempt to find cause and effect in the movements of the stars, and many of the greatest astronomers of medieval times were also astrologers. However, greater understanding of our place in the Universe soon revealed that the planets were enormously distant, and the stars even farther away. Gradually, the idea that they had a major influence on events here on Earth dwindled.

AN EXPANDING VIEW OF THE UNIVERSE

In many ways, the history of astronomy has been a progressive acceptance of humankind's place in the Universe. Over the past 500 years, our ideas have continually changed. The geocentric view of Earth at the centre of a relatively small Solar System enclosed in a spherical shell of fixed stars changed to a heliocentric view of the Sun at the centre of everything, with Earth as one of several planets orbiting it. Then came the realization that the stars were suns like our own, at unimaginable distances, and that our Sun is a fairly dim star on the outskirts of the Milky Way. Finally, in the 20th century, astronomers proved that our galaxy is just one of many, and Einstein and his theory of relativity showed that no place in the Universe is more special than any other.

Coupled with these changes in outlook have come advances in technology. Prehistoric people showed remarkable sophistication in tracking the movements of the Sun, Moon and stars with everything from standing stones to immense temples. Medieval and Renaissance astronomers brought precision to astronomy with the astrolabe and the quadrant, and the revolution in astronomy that followed owed as much to the patient labour of Tycho Brahe in accurately

plotting the paths of the planets around the sky as it did to the glamourous telescopic discoveries of Galileo Galilei. William Herschel's discovery in 1800 of invisible radiation beyond the range of visible light led to the development of spectroscopes that unlocked the chemical composition of the heavens. Meanwhile, photography was eagerly seized upon as a method of recording the light of faint stars beyond the reach of the human eye. In the 20th century, many of the scientists and engineers who brought the world into the Space Age were inspired by the applications for astronomy—the first astronomical instruments launched into space were carried by V-2 rockets left over from the Nazi war effort.

THE CONTINUING ADVENTURE

One might think that, with all the advances made in recent decades, astronomers are now

close to understanding everything about the Universe. But many of their ideas are just working models, and often prove to be fatally flawed when attempting to explain new observations. As technology advances and new mysteries reveal themselves, the adventure begun by our prehistoric ancestors will surely continue for centuries to come.

Ancient astronomy
The prehistoric monument of Stonehenge on Salisbury Plain, England, built circa 3000-1500 B.C., is thought to have been used for religious rituals that were linked to astronomical events.

High-tech astronomy
The Hubble Space Telescope (opposite) has returned the clearest images from the depths of the Universe.

Ancient theories

Nearly every ancient culture seems to have developed its own astronomy and cosmology to explain the Universe around them. Although some of the most fascinating civilizations left no written record to explain their beliefs, others became prolific record-keepers, and passed down observations and theories that continued to affect astronomers' studies for centuries.

Astronomy developed independently on every continent, taking on a variety of forms ranging from simple folk belief to scientific discipline.

EARLY TRADITIONS

The first records of the night sky date back thousands of years. The earliest known map of the Moon, for example, comes from a stone tomb at Knowth, Ireland, and dates to around 3000 B.C. From this time onward, early European people erected enormous stone circles and rows that aligned with the rising and setting of the Sun at different times of year. Although Stonehenge and the many smaller monuments across northern Europe are often described as calendars or astronomical calculators, their sheer scale shows they must have had a far deeper significance for the people who built them. Intriguingly, many of these megalithic monuments are aligned to the important rising and setting points of the Moon during its complex 18.6-year cycle of movements. This shows that the megalith builders were sophisticated skywatchers, capable of keeping records in oral form over many generations.

Several other cultures developed oral traditions based on the night sky. These range from Inuit and Australian aborigine legends to the sophisticated star lists learned by the sailors of the Caroline Islands in the South Pacific, which enabled them to navigate between small islands hundreds of miles apart.

Most Egyptian astronomy was associated with mythology, as one might expect in a culture so dominated by its religion. The sky was seen as the arched body of the goddess Nut, and the Milky Way may have been associated with the River Nile, source of all life and fertility in the midst of the arid desert. Great temples were built with alignments toward the rising Sun, and the constellation of Orion also seems to have been very important in rituals associated with the afterlife. More practically, the Egyptians learned to associate the helical rising of Sirius – when the bright star rises just before the Sun—with the approach of the flood season that would water their crops for the next year.

Later, the civilizations of Mesoamerica and China also developed complex astronomical knowledge. The Maya and Aztecs erected huge temples aligned with the movements of the Sun and Venus, and even built whole cities on astronomical alignments. During the spring and autumn equinoxes the shadows cast by the setting sun on the so-called Castello, or Pyramid of Kukulcán, create a huge snake slithering down the steps.

The Chinese kept extremely accurate astronomical records and learned how to predict eclipses and other celestial events. But both of these alternate astronomies developed in isolation. The origins of modern astronomy lie in the cradle of western civilization – Mesopotamia (modern-day Iraq).

THE IMPORTANCE OF RECORDS

In Mesopotamia, a complex mathematical astronomy had developed by around 1800 B.C. that affected all the civilizations that followed. The Mesopotamians were astrologers, although they had sophisticated beliefs about the relationship between the skies and events on Earth. They thought that history was cyclic

Mayan astronomy
The pyramid of Kukulcán is part of the Mayan city of Chichén Itzá in Yucatan. Each of its four staircases has 91 steps, and including the step on the top platform they add up to to 365 – the number of days in the solar year.

and that movements in the heavens simply followed the same cycle as events on Earth. There was no direct cause and effect between the two, but movements in the sky offered omens about possible events on Earth.

The shapes and descriptions of many of our modern constellations originate with the Mesopotamians, and they left a legacy of accurate record-keeping that astronomers can use even today. For example, the discovery of the

earliest recorded sighting of Halley's Comet – from Babylonian records dated at 164 B.C. – helped to establish its orbit with unprecedented accuracy.

THE UNIVERSE OF THE MATHEMATICIANS

From Mesopotamia, the baton of astronomical progress passed to classical Greece. From around 400 B.C. onward, observation met philosophy as the first Greek scientists tried to construct models of the Universe. Most classical astronomy was based on geometry. Using relatively simple principles, the Greeks were able to make a series of important advances, giving a scale to the Universe for the first time. They measured the size of Earth and made the first estimates of the distance to the Sun and Moon, which also provided them with approximate sizes for both. This convinced one school, led by Aristarchus, that the Sun, not Earth, was the centre of the Universe.

THE CENTRE OF THE UNIVERSE

The vast majority of ancient astronomers, however, were convinced that Earth was the centre of everything – a geocentric Universe. At first glance, this seems self-evident, but problems soon arise in trying to explain the motions of the planets according to this model. To add to their problems, ancient Greek astronomers were philosophically attached to the idea that everything moved in perfect circular orbits.

In the second century A.D., the Greek-Egyptian astronomer and geographer Ptolemy finally came up with a geocentric model of the Universe that matched reasonably well with observation. It was hardly questioned for more than a millennium. Ptolemy's cosmology only survived for so long because, at the time he was writing, western civilization was on the brink of a long decline from which it would not emerge until the Renaissance.

Geocentric Universe
Early astronomers and mathematicians could not bring themselves to believe that Earth was not the centre of the Universe, and this picture (left) remained current until the times of Copernicus and Galileo.

PTOLEMY

Claudius Ptolemaeus, better known as Ptolemy, was a Greek-Egyptian astronomer working in Alexandria - then part of the Roman Empire. He left a lasting influence on history in the form of two great works - the *Geography* and the *Almagest*. The *Almagest* was the culmination of classical astronomy. Ptolemy proposed that the Sun, Moon and planets moved round the Earth in perfect circles. However, to explain obvious problems, such as the way Mercury and Venus stayed near the Sun, and the occasional backward or retrograde movements of Mars, Ptolemy added small extra circles, called epicycles, to the orbits of the planets. Every time a planet caught up with an epicycle, it would have to loop around it before carrying on along its main orbit. Although best remembered for his geocentric Universe, Ptolemy also compiled a definitive list of 48 constellations that we still use - with additions - to this day.

Almost 1,500 years passed between Ptolemy's *Almagest* and the revolution that finally broke astronomy free of the old geocentric ideas. During this time, while western Europe languished in the so-called Dark Ages, the science was nurtured in the Arab world, where the art of observation reached new heights.

ASTRONOMY PASSES TO ISLAM

While progress in Europe stalled, some advances in astronomy continued in the East, as Muslim astronomers adopted and improved upon the writings of the ancients. Most of the individual names given to bright stars today are derived from Arabic. Islamic astronomy was driven by certain religious requirements—for instance, daily prayers had to be precisely timed according to the position of the Sun and always directed toward the holy city of Mecca. In addition, the exact time of the New Moon had to be known, since the Muslim calendar is based around lunar months rather than the movement of the Sun.

In Europe, too, the astronomy that survived was driven by religion – most importantly, the need to calculate Easter, which was linked to the first Full Moon after the spring equinox. Although frowned upon by religious authorities, astrology also flourished. Around the start of the second millennium, translations of ancient Greek astronomical works started to filter back into Europe and the astrolabe, a sophisticated astronomical instrument for measurement and calculation, was introduced from the Arab world. A slow recovery in western astronomy now began – the great Gothic cathedrals of the 12th century onward incorporate astronomical features, as do some of the greatest works of medieval literature. By the 13th and 14th centuries, Islamic science and the founding of the first universities reintroduced the idea of empiricism – the development of theory from observation, not philosophy.

This renewal of empirical science, coupled with new technology, was bound to reveal the major problems in the geocentric Universe theory. But, ironically, the person who takes most credit for introducing the picture of a heliocentric Universe, the Polish priest Nicolaus Copernicus (1473–1543), actually developed his system because of philosophical objections to Ptolemy's theory.

PUTTING THE SUN CENTRE STAGE

Copernicus worked in the early 16th century, by which time it was well known that some of the Greeks had thought that Earth moved around the Sun. Copernicus objected to Ptolemy's world picture chiefly because the geocentric theory required that the planets should alter their speed as they moved around their orbits and epicycles. Copernicus and many others felt that motion around a circle should be uniform, and he was able to show that a heliocentric Universe could meet this condition. However, his theory was hardly any simpler than Ptolemy's. It still involved complex patterns of epicycles to explain deviations from the predictable circular orbits.

Copernicus knew that his ideas contradicted the teachings of the Catholic church and was careful to present them only as theories. Although he circulated his ideas as early as 1513, his great summary of them, *On the Revolutions of the Heavenly Spheres*, had to be printed in the Protestant city of Nuremberg and was not published until 1543, the year of his death. At first, Copernicus' theories made

remarkably little impact. It was only in the later 1500s that advances in observation conclusively proved Ptolemy wrong.

SUPERNOVAE AND COMETS

In 1572, a new star – a supernova – appeared in the constellation Cassiopeia. It was observed by a young Danish astronomer, Tycho Brahe (1546–1601), who realized that it offered important evidence that the stars were not some kind of fixed, celestial backdrop but were real objects, susceptible to change.

Brahe was the greatest astronomer of his time. Working a generation before the invention of the telescope, he devised huge instruments that brought unprecedented accuracy to his measurements of stars and planets. In 1577, a great comet appeared in the skies over Europe. Bright enough to be seen in daylight, it made a huge impression on all who saw it, including Brahe. He used the parallax method to measure the distance to the comet, proving that it was not – as astronomers had previously believed – an atmospheric phenomenon, but was far from Earth, in the realm of the planets. The path the comet took meant it must be crossing the orbits of the planets – the first conclusive evidence that the universe did not work in perfect circles.

Although Brahe marshalled most of the evidence that finally brought the geocentric system crashing down, he could not bring himself to remove Earth from its position at the centre of everything. Instead, he devised a compromised Tychonic system in which the Sun and Moon went around the Earth but all the other worlds moved around the Sun.

HEAVENLY LAWS

The final overthrow of the geocentric Universe came in the early 1600s. It was the work of Brahe's German student and successor, Johannes Kepler (1571–1630) and the Italian astronomer Galileo Galilei (1564–1642), who made many of the first telescopic observations

Kepler was a brilliant mathematical astronomer. He took on the gargantuan task of reanalysing Brahe's records and concluded that Earth was moving around the Sun, and that the planets occupied elliptical, not circular, orbits obeying a simple set of rules – Kepler's three laws of planetary motion. He published his discoveries in two great works – the *Astronomia*

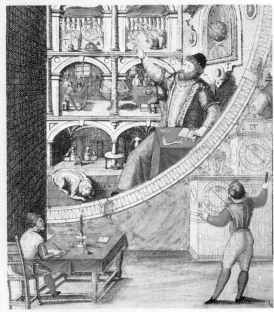

Nova in 1608 and the *Harmonice Mundi* (*Harmony of the World*) in 1619.

Telescopes, invented by the Dutch lensmaker Hans Lippershey, came into use around 1608 and were immediately adopted by several astronomers, including Galileo, who announced his discovery of moons orbiting Jupiter in 1610. Realizing that not everything in the Universe moved around Earth, Galileo became an ardent supporter of the Copernican model. By the time of his death this model was accepted by most European astronomers.

GALILEO GALILEI

Galileo Galilei was professor of mathematics at the University of Pisa in Italy, where he is famously supposed to have proved that all objects fall at the same rate under gravity by dropping cannonballs of different weights from the Leaning Tower. He began making telescopes and published his discoveries of the Galilean moons, the phases of Venus, and the stars of the Milky Way in his 1610 book *The Starry Messenger*. His discoveries convinced him that Copernicus was right, but his teaching of the heliocentric Universe as fact rather than theory led to a confrontation with the Catholic church. Following his *Dialogues on the Two Chief Systems of the World* (1632), he was called before the Inquisition, tried for heresy, and kept under house arrest for his last years. For his sacrifice, Galileo became a hero to later generations of scientists.

Enlightenment astronomy

Astronomy blossomed as a science during the 17th and 18th centuries. Kepler's solution to the problem of planetary orbits had provided the first real glimpse of how the Universe worked. Newton's discovery of gravitation went on to explain why it worked, while increasingly powerful telescopes began to reveal new moons, planets, stars, and nebulae.

THE REAL UNIVERSE

Despite opposition from the Catholic church, the heliocentric view of the Universe was thoroughly established by the mid-17th century. As telescope apertures increased, so did evidence that the other planets in our Solar System were worlds like our own. Saturn revealed its rings, Jupiter its cloud-belts and storms, and Mars its dark surface features. New satellites orbiting other planets were discovered, joining the Galilean moons – most notably Saturn's large companion, Titan.

Meanwhile, many of the world's finest minds, including French philosopher René Descartes (1596–1650) and English physicist Robert Hooke (1635–1703), were trying to understand the forces that kept the Universe running. The old Ptolemaic idea had been that the planets were fixed to transparent crystal spheres, and driven by a sort of celestial clockwork. But Tycho Brahe had proved that the comet of 1577 was cutting across the supposed spheres of the planets, so they could not actually be there. The scale of the planetary orbits was also becoming apparent now that Earth had been put in its proper place around the Sun.

The problem was solved by the brilliant English mathematician and scientist Isaac Newton (1642–1727) who, according to legend, watched an apple fall to the ground and realized that the same force must also be acting on the Moon. Since the Moon does not fall to the ground, it must be attempting to move away from Earth with just enough acceleration to counteract gravity. He also realized that if the Earth was exerting a force on the Moon, then the Moon must also be exerting a force on the Earth.

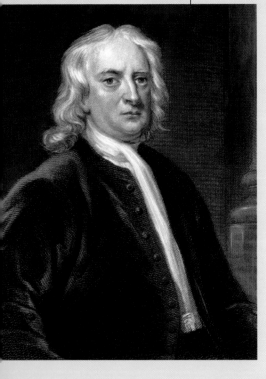

Ahead of his time
Newton developed his Law of Universal Gravitation as early as 1665, yet another hundred years passed before his theories became fully accepted by the world of science.

EDMOND HALLEY

British astronomer Edmond Halley (1656-1742) is best known for the comet that bears his name. His most important contribution to science was probably his successful prediction of Comet Halley's return in 1758. The event proved that Newton's theories on the Law of Universal Gravitation, which Halley had done so much to bring to public attention, were indeed correct. Halley made many other discoveries, however. As commander of a Royal Navy warship on a scientific expedition to the southern hemisphere, he studied meteorology and oceanography, mapped the southern skies and was among the first to observe a transit of Mercury across the face of the Sun. In 1710, Halley's studies of the star catalogues compiled by Ptolemy in the 2nd century A.D. led him to the conclusion that some of the stars were moving—a discovery he later confirmed with observations of the bright stars Sirius, Arcturus and Procyon.

Newton used Kepler's laws, which summarized the motion of objects in orbit without explaining them, to derive a mathematical expression for the gravity between two objects – his Law of Universal Gravitation. He also described the behaviour of objects subject to forces in three Laws of Mechanics. This work went unannounced for years until it was finally published in 1687, with the encouragement and financing of Newton's friend Edmond Halley.

However, Newton's theory was not universally welcomed. In France, Descartes' theory that everything moved because it was buffeted by an interplanetary medium survived for several decades. The final triumph of Newtonianism did not come until 1758, when a comet previously seen in 1607 and 1682 reappeared just as predicted by Halley using Newton's formulae.

MAPPING THE SKIES

Both Halley and Newton were members of the Royal Society – one of the first scientific societies, founded in London in 1660. The reign of King Charles II was a good one for British science in general and astronomy in particular, and saw the foundation of the Royal Observatory at Greenwich. In an age where exploration was rapidly giving way to worldwide trade, astronomy was the principal tool of navigation. Astronomers such as Halley travelled to the southern hemisphere and catalogued the stars for practical, as well as scientific reasons.

This was the era of the great star indexers – Johann Bayer (1572–1625) had devised his system of letters for listing a constellation's stars in the early 1600s, just before the invention of the telescope revealed thousands more stars to be catalogued. Charles Messier (1730–1817) compiled the first catalogue of nebulae and, with the opening up of the southern skies, new constellations sprang up like mushrooms. John Goodricke (1764–1786) and others also identified the first variable stars.

Astronomers were slow to see that the telescope could also be used to give new precision to the measurements of stellar positions. One of the first to realize this was Dutch astronomer Christiaan Huygens (1629–1695). He developed the eyepiece micrometer, consisting of movable wire crosshairs to measure the separation of two objects within a telescope's field of view. Coupled with more sophisticated mountings that enabled the direction of the telescope to be accurately recorded, the precision of star positions increased to just a few minutes of arc – $^{1}/_{60}$ths of a degree – and, ultimately, to just seconds of arc – $^{1}/_{3,600}$ths of a degree.

PRECISION MEASUREMENTS

This new accuracy finally allowed astronomers to measure the fine movements of the stars – proper motions – and the orbits of double stars around each other. It also allowed the orbits of planets to be determined even more precisely, leading to the discovery of minute variations from their predicted positions.

These perturbations were eventually explained by the French mathematician Pierre Simon Laplace (1749–1827), who showed that they were caused by the gravity of the planets affecting one another. Laplace published his discoveries in his 1796 *Exposition du Système du Monde* (*Explanation of the World System*), which also presented the first scientific model of the origin of the Universe. His theory – that the Solar System had formed from a large gas cloud – was correct in spirit, though wrong in detail.

Although Laplace's development of celestial mechanics was key to the later discoveries of the outer planets Neptune and Pluto, the first of the new planets, Uranus, was discovered by chance. In 1781, German-born astronomer William Herschel (1738–1822) was studying the skies from his back garden in Bath, England, when he spotted an object that he had previously dismissed as a comet. Observation over several nights revealed that it moved too slowly and must be a new planet beyond Saturn. Herschel built the finest telescopes of the day, and used them to make advanced studies. Ironically, Uranus is a borderline naked-eye object that had been seen and catalogued by less perceptive observers several times before.

Legacy of the Enlightenment The Royal Greenwich Observatory was founded in 1675, after a similar institution opened in Paris. It continued to operate as an observatory until 1998, making it the oldest scientific institution in Britain at the time.

The arrival of astrophysics

The 19th century saw the transition from positional astronomy – the study of the motions of celestial bodies based on the accurate measurement of their positions – to astrophysics, the study of the objects themselves based on both the light they emit and theoretical models.

THE LIGHT FROM STARS

The key to this change was spectroscopy, first seen when Isaac Newton split sunlight through a prism into many colours. The English chemist William Hyde Wollaston (1766–1828) refined the process by passing the Sun's light through a narrow slit before splitting it into a spectrum. The narrow light source stopped the smearing of the Sun's image across the spectrum, and revealed fine features in the light – several dark lines indicating colours where the Sun was apparently not producing light. Further refined by the German optician Joseph von Fraunhofer (1787–1826) in 1821, the solar spectrum proved to be covered with hundreds of dark lines.

Fraunhofer noted that one of the most prominent dark lines seemed to coincide with a type of yellow light produced in flames. But it was not until 1859 that the German scientists

Robert Bunsen (1811–1899) and Gustav Kirchoff (1824–1887) showed a connection between certain groups of lines and specific elements. Kirchoff also showed how a hot gas that normally emitted light with precise colours would absorb the same colours when lit from behind with a source of white light containing many colours combined. The same thing clearly happened in the Sun – a continuous spectrum of light from the photosphere was passing through a hot outer atmosphere that absorbed certain frequencies of light, producing the dark lines.

Experiments conducted at eclipses showed that, with the photosphere hidden, the Sun's outer layers produced an emission spectrum of the same colours it normally absorbed. These discoveries allowed astronomers to work out the chemical composition of the Sun, and led to the discovery of a new element, helium, unknown at that time on Earth.

DISTANCES TO STARS

The mid-19th century also saw the first accurate measurement of interstellar distances. Galileo had worked out the principle of using parallax from the Earth's movement to measure the distance to stars, but the first accurate measurement was not made until 1838, when the German astronomer Friedrich Bessel (1784–1846) successfully measured the shift in position of the faint star 61 Cygni, and proved that it lay about 10.3 light-years from Earth.

By then, most astronomers suspected that the stars were suns like our own, and Bessel and others began to build up a map of our stellar neighbourhood based on parallax. This revealed that the stars had intrinsic differences in brightness as well as colour. Some were fainter than the Sun, while others were thousands of times brighter.

OUR GALAXY AND OTHERS

Once the distances to some stars were known, astronomers could begin to understand the patterns they formed in the sky. William Herschel was the first person to realize

the significance of the Milky Way and attempt to draw a map of our galaxy. He saw that it must be a flattened disc, but had no way of knowing about its spiral arms. At the time, no one connected our galaxy with the patches of misty light catalogued by Messier and others. Herschel, for example, thought that they formed a halo around the galaxy.

Astronomers were also uncertain of the nature of these nebulae. Were they clusters of of countless stars. Others were clearly made of gas – like hot gases in the laboratory, they showed an emission spectrum with just a few bright coloured lines.

Using data compiled by Henry Draper at Harvard, Henry Norris Russell (1877–1957) and Ejnar Hertzsprung (1873–1967) in the early 1900s plotted the first Hertzsprung-Russell diagrams, comparing the spectral types of stars with their intrinsic luminosities derived from

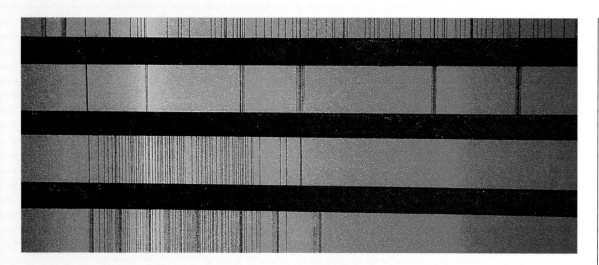

stars so distant that they could not be resolved, or really just clouds of gas? The first person to successfully distinguish between the types of nebulae by spectroscopy was the British amateur astronomer William Huggins (1824–1910). To do this, he made extensive use of a recent invention – photography.

THE PHOTOGRAPHIC REVOLUTION
Although the principle of spectroscopy is, in theory, applicable to all stars and other objects, even the largest telescope could not collect enough light from a star to split it into a useable spectrum. Long-exposure photography allowed light to be collected from stars for several minutes or even hours, capturing far more light than could be seen by an observer. It also allowed accurate measurements of spectral lines, and formed a permanent record for direct comparison. In this way, astronomers discovered that stars of different colours also tended to have different chemicals in their atmospheres.

Spectroscopy allowed Huggins to prove that nebulae were of two distinct types. Some showed a starlike spectrum, with a continuum of light crossed by dark lines, indicating they were made

parallax measurements. Photography also allowed astronomers to make some of the first truly accurate maps of other planets. It enabled discoveries of things far beyond the limits of the naked eye – such as new moons, faint structure in gaseous nebulae and millions of new stars.

Stellar spectra
US astronomer Henry Draper (1837-1882) was among the first to use photography to record stellar spectra, and the first great catalogue of spectra still bears his name. Astronomers such as Cecilia Payne-Gaposchkin (1900-1979) worked out the actual ratios of elements in stars, showing that they are largely made of hydrogen.

GEORGE ELLERY HALE

George Ellery Hale (1868-1938) was a prominent figure in the transformation of astronomy. As a student at the Massachusetts Institute of Technology, he devised the spectroheliograph, so that the Sun's outer atmosphere could be observed without an eclipse. Appointed professor of astrophysics at the University of Chicago in 1892, he founded the Astrophysical Journal (having coined the word "astrophysics") and raised funds for the construction of several large tele-scopes. In 1908, his 150-cm (60-inch) reflector was completed, followed in 1917 by a 2.5-m (100-inch) telescope. Hale also supported the founding of the American Astronomical Society. Before his death in 1938, he had secured funding from the Rockefeller Foundation for the 5-m (200-inch) telescope on Palomar Mountain, California.

Einstein and beyond

At the turn of the 20th century, astronomy was on the verge of a breakthrough in its understanding of the cosmos. New discoveries that followed, changing our conception of space-time and atomic physics, would finally offer solutions to some of the Universe's greatest riddles.

SHEDDING LIGHT ON LIGHT

In the late 19th century two questions in physics became pressing: the nature of light, and the nature of atoms. It had long been realized that light was a wave, but a wave needs a medium to travel through, and the medium for light, called the ether, had never been detected. When experiments by Albert Einstein (1879–1955) and others showed that light also behaved like a particle in some situations, it became clear that the ether did not exist, and light somehow combined the properties of both a particle and a wave. One of the strangest discoveries was that light travels at a constant speed regardless of the motion of the light's source.

Radioactivity was another major pre-occupation of early twentieth-century physicists. In a few decades, scientists discovered all the major subatomic particles – protons, neutrons and electrons. They also realized that during radioactive decay, atoms of one element lose subatomic particles and transform themselves into atoms of a different element. The Danish physicist Niels Bohr (1885–1962) showed how different elements would emit or absorb light of specific energy and colour as the electrons jumped between fixed energy levels within each atom. This finally explained the origin of the absorption lines in stellar spectra, and the emission lines in nebular spectra.

Astrophysics was also developing rapidly. Cecilia Payne-Gaposchkin's studies of stellar spectra showed that the atmospheres of stars were overwhelmingly composed of hydrogen and helium. The British astronomer Arthur Eddington (1882–1944) studied binary stars to work out their orbits and find their masses from Newton's law of gravitation. In 1924 he announced that there was a clear link between mass and luminosity for stars on the Hertzsprung-Russell diagram's main sequence.

THE ENERGY OF THE STARS

Eddington was also the first person to suggest that the Sun's energy might come from nuclear fusion. He reasoned in 1927 that, deep in the Sun's core, nuclei of hydrogen might be forced together under extremes of temperature and pressure to form helium. In 1938 the German physicist Hans Bethe (1906–) proved Eddington right by showing how the fusion of light elements could release even more energy than the radioactive decay (fission) of heavier ones.

With the findings of the atomic physicists, it was possible to work out an accurate lifespan for the Sun – by calculating the amount of energy theoretically available in its core (using Einstein's $E = mc^2$), and how much power the Sun generates every second. The resulting lifespan – roughly 10 billion years on the main

Eclipses and relativity
In 1919 Arthur Eddington observed that stars close to the Sun in an eclipse appear to have moved from their normal positions. This showed that Einstein's theory was right, and the Sun's mass was deflecting light passing close to it.

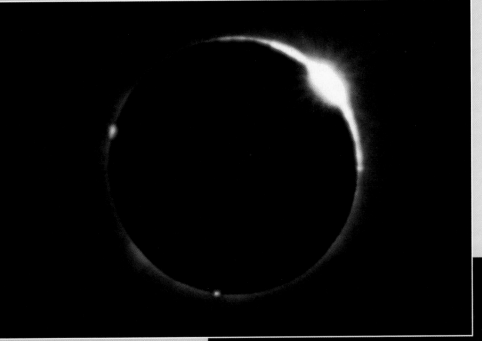

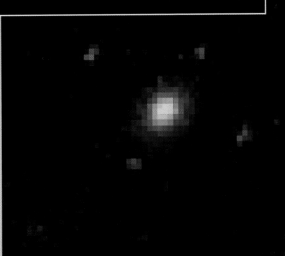

Gravitational lensing
This Hubble Space Telescope image shows relativity at work. The bright galaxy in the centre is bending space close to it, creating an Einstein cross - four distorted images of the same galaxy behind it.

EDWIN HUBBLE

A central figure in twentieth-century astronomy, Edwin Hubble (1889-1953) trained as a lawyer before returning to his first love of astronomy. In 1918, he went to the Mount Wilson Observatory in California and his studies there led to the realization for the first time that our galaxy, far from being the entire Universe, is only one of hundreds of thousands of star systems. Hubble studied Cepheid variable stars and found tell-tale Cepheids in the Andromeda Galaxy, proving it lay over 2 million light-years away. Even more significantly, in 1929 he went on to discover that light from these other galaxies was redshifted, indicating that they were moving away from us with surprising speed, and that this speed increased the further away a galaxy was – in other words, the Universe was expanding.

sequence – fit in well with the accepted age of the Solar System, and Eddington soon extended his model to give ages for other stars. He realized that the distribution of stars in the H-R diagram could indicate a process of evolution, with stars spending more or less time in different stages.

EXTREME STARS

Studies of binary systems also revealed stars that did not conform to an expected pattern. The massive, but incredibly small white dwarfs found in some systems were the first of several types of extreme stars whose discovery owed as much to atomic physics as to actual observation. In 1928 Indian physicist Subramanyan Chandrasekhar (1910–1995) suggested that white dwarfs might be collapsed stars whose weight was supported by an internal pressure between their atoms.

In 1932 the Russian physicist Lev Landau (1908–1968) hypothesized that, above the Chandrasekhar limit of 1.4 solar masses, the force of gravity would grow so great that a star's protons and electrons would combine into neutrons, and it would collapse even further to a neutron star – with a diameter just a few kilometres across and supported by the pressure between its neutrons alone. This was confirmed by the discovery of the first pulsar in 1967.

Similarly, the idea of black holes – objects with gravity so strong that light cannot escape

them – was established in the late 18th century. But it was only with the discovery of violent X-ray sources and distant quasars in the 1960s that astronomers had to call on black holes to explain real observations.

OUR PLACE IN THE UNIVERSE

Another breakthrough took place in 1923 when Edwin Hubble proved that our galaxy was just one of many, separated by millions of light-years from each other. He later showed that the distance between the galaxies was increasing, and therefore, the Universe was expanding.

Hubble's discovery fired a new debate about the origins of the Universe – something astronomers could not even have considered properly just a few decades earlier. Previously, many scientists had thought of the Universe as eternal and unchanging, but Hubble's law allowed the expansion to be traced back to a single point in space and time. Russian-born US cosmologist George Gamow (1904–1968) was among the first to suggest an explosive origin for the Universe, although the term Big Bang was coined by one of the idea's fiercest opponents, Fred Hoyle (1915–). Many others also objected to the theory, proposing instead a steady-state cosmos – with new matter continuously created to fill space as the Universe expanded – or an oscillating Universe alternating between expansion and contraction. But Gamow predicted a test for the Big Bang – the afterglow of the explosion should still be detectable.

RADIO ASTRONOMY

Gamow's predicted cosmic background radiation was finally discovered in 1964 by the new science of radio astronomy. The first person to discover radio signals coming from space had been the US engineer Karl Jansky (1905–1950), and in the 1930s another US engineer, Grote Reber (1911–), developed the first dish-shaped radio telescope to map the sky. However, radio astronomy came into its own in the 1950s when several British scientists who had worked on radar during World War II turned their attention to its potential. Their greatest achievement was the 76-m (250-ft) steerable radio telescope completed at Jodrell Bank, England in 1957. This was to be the prototype of many other radio telescopes, and paved the way for astronomy to enter the Space Age.

Albert Einstein
Einstein's Special Theory of Relativity (1905) showed how the constant speed of light created strange phenomena in objects moving close to that speed. In the General Theory (1915) he went on to explain the nature of space-time, and how gravity, light and time are interlinked.

It was not until the 1950s that astronomers were able to look beyond Earth's atmosphere and view the Universe across the range of the electromagnetic spectrum. The first bold experiments began in 1912, when the US physicist Victor Franz Hess (1883–1964) carried out a series of radiation measurements from a hot-air balloon at an altitude of 5 km (3 miles). He found that radiation actually increased above a certain altitude and realized that Earth was being bombarded with high-energy radiation from space.

SPUTNIKS IN SPACE

The Space Age officially began in October 1957 with Sputnik 1, the first artificial satellite in orbit. But long before this, astronomers were using the new rocket technology to get a glimpse of the Universe beyond our atmosphere.

Despite the rocket-borne discoveries, both the Soviet and US space efforts focused on manned spaceflight, the race to the Moon and sending probes to the planets. While some of these vehicles carried simple radiation detectors, most observatories were specially adapted aircraft armed with a battery of different radiation detectors and were designed to fly high in the upper atmosphere – exposed to nearly all of the different radiations from space.

THE FIRST ORBITING HIGH-ENERGY SATELLITES

The first dedicated astronomy satellites were NASA's Orbiting Astronomical Observatories

Space shuttle launch NASA's reusable space shuttle has revolutionized space-based astronomy, acting as a launch vehicle for many satellites and as an orbiting observatory.

(OAOs), launched in the late 1960s. The OAOs made the first map of the whole sky in ultraviolet, and collected ultraviolet spectra from orbit for the first time. European and Soviet satellites also made ultraviolet observations from orbit from the early 1970s. The most effective of these missions was the International Ultraviolet Explorer (IUE), launched in 1978, which functioned for more than 10 years. Since then, the Extreme Ultraviolet Explorer (EUVE) has searched for the hottest ultraviolet objects, and the Solar and Heliospheric Observatory (SOHO) has studied the Sun's outer atmosphere.

Uhuru (SAS-1), the first X-ray detector in orbit, had very poor resolution and could only detect the approximate direction of X-ray sources in the heavens. In the late 1970s, astronomers designed the grazing incidence telescope – a series of nested metal reflectors from which the X-rays ricocheted and were brought to a focus. The first satellite to carry this type of telescope was the High Energy Astrophysical Observatory (HEAO-2), also known as Einstein. This was followed by the European EXOSAT project in the early 1980s, the German-led ROSAT in 1990, NASA's Chandra X-ray observatory in 1999, and the European X-ray Multi-Mirror telescope (XMM) in 2000.

SAS-2, which followed Uhuru into orbit in 1972, was the first gamma-ray observatory in orbit. Gamma-ray astronomy began by accident in the late 1960s, when US Vela satellites launched to detect gamma rays from illegal nuclear weapons tests instead discovered intense blasts of high-energy rays coming from deep space – the gamma-ray bursters.

The high penetrating power of gamma rays means that they cannot be focused. Instead, they are detected using scintillation counters which force the rays to leave a trail of sparks as they pass through metal plates. Or they are found with coded masks, which use a shaped mask to absorb the rays and cast a shadow onto a detector. SAS-2 was followed by the European COS-B satellite, which detected the first gamma rays coming from a pulsar. More recently, NASA launched the Compton Gamma Ray Observatory (CGRO), which operated from 1991 to 2000. This enormous satellite, weighing more than 15 tonnes, made many important discoveries, proving for the first time that gamma-ray bursters were well beyond our galaxy. It also including comets, protoplanetary discs around nearby stars, and unusually bright infrared galaxies. IRAS was followed by the Infrared Space Observatory (1995–1998), which studied the interiors of star-forming nebulae, colliding galaxies, and interstellar gas clouds. The Cosmic Background Explorer (COBE), was launched in 1989 and was designed specifically to map faint cool radiation left over from the Big Bang.

THE HUBBLE SPACE TELESCOPE

The idea of a space-based optical observatory was first suggested by US astrophysicist Lyman Spitzer (1914–1997) in 1946 but NASA did not begin serious work on such a project until the late 1970s. An orbiting optical telescope avoids the distortion caused by Earth's atmosphere, allowing astronomers to see more clearly into the depths of the observable Universe. The project was eventually named the Hubble Space Telescope (HST), since its main goal would be to establish a precise value for the Hubble constant – the Universe's rate of expansion. Launched in 1990, it has been a spectacular success since its faulty mission was repaired in 1993, returning hundreds of images and allowing us to see the Universe with a new clarity. However, the HST was only designed for a 15-year operating life, so planning is now underway for its successor, the Next Generation Space Telescope (NGST), to be launched in 2009. This will probably be a multiple-mirror telescope with an overall diameter of 8 m (26 ft) capable of probing to the very edges of the Universe.

V-2 rocket In 1946 the US Naval Research Laboratory launched a German V-2 rocket, captured at the end of World War II, to a height of 80 km (50 miles), where it obtained the first full spectrum of the Sun from above the ozone layer.

provided information about soft gamma repeaters – believed to be caused by starquakes in the crust of neutron stars.

INFRARED ASTRONOMY FROM SPACE

Infrared astronomy presents a different problem. Although some infrared wavelengths can be seen by mountaintop telescopes, most are blocked by water vapour in Earth's atmosphere. Satellite telescopes need to be constantly cooled in order to stop their own heat from interfering with the faint infrared from the stars. They are cooled with an extremely cold liquid such as liquid helium but even this will gradually heat up and, in space, can never be replenished.

However, the few infrared telescopes that have been put in orbit have been very successful. The first was the Infrared Astronomy Satellite (IRAS), launched in 1983. IRAS functioned for 300 days and surveyed 98 per cent of the sky at various infrared wavelengths, detecting a quarter of a million sources of radiation

A clearer view The Hubble Space Telescope is untroubled by the perturbations of Earth's atmosphere, enabling it to capture images of astounding clarity in the visible spectrum.

Probes to the planets

Pioneer's message
Pioneer's message
Pioneers 10 and 11 were the first probes to leave the planets behind and begin the long journey into interstellar space. They each carried a plaque (right) engraved with a message from Earth - images of humans, and a map of how to find our planet.

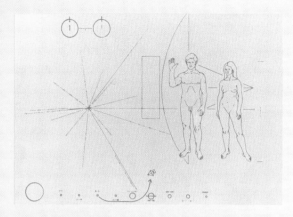

Once the first satellites had been launched, the exploration of our Solar System was an obvious next step. Journeys to other planets had long been the subject of science fiction, but no one could have imagined the advances in technology that would send robot probes, instead of astronauts, into the depths of space.

THE RACE BEGINS

The first probes to go beyond the immediate vicinity of Earth were aimed at the Moon. At the dawn of the Space Age in the late 1950s, the Soviet Union had the more powerful rockets and were able to launch probes long before NASA. These rockets were capable of achieving speeds close to the escape velocity of 11.2 km per second (7 mps) needed to break free of Earth's gravity altogether.

The first probes were aimed at crashing into the Moon and sending back images up to the moment of impact. Lunik 2 made the first successful impact in 1959 and later that year, Lunik 3 swung behind the Moon, sending back pictures of its far side for the first time. More powerful rocket engines enabled the Soviet Union to launch probes to Venus and Mars. However, they ran into repeated technical problems and by the late 1960s the US had caught up.

The first US probes were the Ranger missions. Ranger 7 crash-landed in July 1964 and helped answer a major debate – whether the Moon's

craters were volcanic or caused by asteroid impacts. By photographing craters of all sizes, right down to a few inches across, it proved that impacts were the cause.

This discovery left scientists wondering whether the Moon's surface would be stable enough to support the weight of a vehicle - the only way to test the lunar soil was to make a soft landing. The Soviets made the first breakthrough, landing Luna 9 successfully in January 1966. Since the probe did not sink, subsequent landings were deemed possible.

Luna 9 also pioneered a new approach to spaceprobes – splitting a vehicle in two. In this case, the soft lander simply dropped from a mother ship shortly before the larger vehicle smashed into the lunar surface. Later on, landers were fitted with retro-rockets that would fire to slow their descent. More recent missions

The surface of Venus
The Magellan Venus probe mapped the planet with a high-powered radar system. Using modern computing techniques, astronomers are able to turn radar data, or photographs taken from several angles in space, into three-dimensional maps of a planetary surface.

have used a main vehicle that stays in orbit above the surface to conduct photographic surveys, while a smaller lander sends data from the ground.

BRAVE NEW WORLDS

In the early days of the Space Race, simply reaching other worlds was an achievement. NASA launched their Mariner probes, which opened up the inner Solar System, in the 1960s and early 1970s. Venus was the first target and in 1963, Mariner 2 revealed that the planet was cloaked in a choking atmosphere. Later, the Soviet Venera landers successfully landed on the surface but could only send back information for a few minutes before being destroyed.

The first Mariner probes to Mars passed over the planet's rugged and forbidding southern hemisphere, leading most scientists to dismiss Mars as a geologically dead world. Yet when Mariner 9 successfully went into orbit around the planet in 1971, it revealed the huge volcanoes, deep chasms and possible river channels of the planet's northern regions.

This was the first probe to enter orbit around another planet – an option not available until rockets became powerful enough slow the probe down and drop it into orbit. Today, almost every mission is designed to orbit its target – whether it be a planet, moon, comet, or asteroid – for an extended period, using new technologies such as radar mappers to probe its secrets.

The final Mariner probe, Mariner 10, is the only probe to have visited Mercury – a particular challenge because the planet travels so quickly. Any probe must gain a great deal of speed if it is to make anything more than a fleeting encounter. Mariner 10 used the gravity of Venus to put it into an orbit that gave it three close encounters with Mercury in the space of a year.

THE OUTER LIMITS

NASA knew that a chance alignment of planetary orbits around 1980 would make it possible to launch a probe on a grand tour of all the outer planets, swinging around each in turn and using its gravity to pick up speed and change direction.

The Voyager missions were launched in 1977, equipped with an arsenal of scientific instruments and the best available cameras.

Flying past Jupiter in 1979, they discovered the planet's ring and photographed the varied surfaces of the Galilean moons for the first time. At Saturn, they photographed details of the planet's rings and the storms in its atmosphere. Voyager 1 then flew past the mysterious giant moon Titan, while Voyager 2 used Saturn's gravity to put it on course for Uranus and finally, Neptune. It is still the only probe to have visited these outer gas-giant planets, while Pluto remains the only planet not visited by any spaceprobes.

As the planetary alignment drifted apart, NASA returned to the idea of single missions targeted toward individual planets and minor members of the Solar System. Galileo went into orbit around Jupiter in 1995, and was still functioning in 2001. The Cassini mission to Saturn will arrive there in 2004, and carries with it a probe designed to land on Titan. Meanwhile, Mars remains the focus of intense exploration and speculation. Plans are also being made for a Mercury orbiter and a dedicated mission to Jupiter's icy moon Europa.

Cassini
The Cassini probe will arrive at Saturn in 2004. It carries a lander that will plunge into the atmosphere of the giant moon Titan. Once this is deployed, the probe will go into orbit for several years, studying the giant planet, its rings, and its many moons.

CARL SAGAN

Best known as a promoter of astronomy, Carl Sagan (1934-1996) was also a mission specialist for several NASA space-probes, including the Mariner Venus missions, the Viking Mars probes, the Pioneers, Voyagers and Galileo. He was the first person to establish the surface temperature of Venus and was a specialist in planetary atmospheres. He was also a keen supporter of the Search for Extraterrestrial Intelligence (SETI). He wrote many popular science books, the novel *Contact* and made the television series *Cosmos*. However, his most important legacy might ultimately be the messages to other civilizations he ensured were carried by the Pioneer and Voyager probes as they flew out and beyond the Solar System.

Constellations and bright stars

CONSTELLATIONS

NAME	GENITIVE	TRANSLATION	SHORT FORM
Andromeda	Andromedae	Andromeda	And
Antlia	Antliae	The Air Pump	Ant
Apus	Apodis	The Bird of Paradise	Aps
Aquarius	Aquarii	The Water Carrier	Aqr
Aquila	Aquilae	The Eagle	Aql
Ara	Arae	The Altar	Ara
Aries	Arietis	The Ram	Ari
Auriga	Aurigae	The Charioteer	Aur
Boötes	Boötis	The Herdsman	Boo
Caelum	Caeli	The Chisel	Cae
Camelopardalis	Camelopardalis	The Giraffe	Cam
Cancer	Cancri	The Crab	Cnc
Canes Venatici	Canum Venaticorum	The Hunting Dogs	CVn
Canis Major	Canis Majoris	The Great Dog	CMa
Canis Minor	Canis Minoris	The Little Dog	CMi
Capricornus	Capricorni	The Sea Goat	Cap
Carina	Carinae	The Keel	Car
Cassiopeia	Cassiopeiae	Cassiopeia	Cas
Centaurus	Centauri	The Centaur	Cen
Cepheus	Cephei	Cepheus	Cep
Cetus	Ceti	The Whale	Cet
Chamaeleon	Chamaeleontis	The Chameleon	Cha
Circinus	Circini	The Compasses	Cir
Columba	Columbae	The Dove	Col
Coma Berenices	Comae Berenices	Berenice's Hair	Com
Corona Australis	Coronae Australis	The Southern Crown	CrA
Corona Borealis	Coronae Borealis	The Northern Crown	CrB
Corvus	Corvi	The Crow	Crv
Crater	Crateris	The Cup	Crt
Crux	Crucis	The (Southern) Cross	Cru
Cygnus	Cygni	The Swan	Cyg
Delphinus	Delphini	The Dolphin	Del
Dorado	Doradus	The Goldfish	Dor

CONSTELLATIONS

NAME	GENITIVE	TRANSLATION	SHORT FORM
Draco	Draconis	The Dragon	Dra
Equuleus	Equulei	The Foal	Equ
Eridanus	Eridani	The River	Eri
Fornax	Fornacis	The Furnace	For
Gemini	Geminorum	The Twins	Gem
Grus	Gruis	The Crane	Gru
Hercules	Herculis	Hercules	Her
Horologium	Horologii	The Clock	Hor
Hydra	Hydrae	The Water Snake	Hya
Hydrus	Hydri	The Little Water Snake	Hyi
Indus	Indi	The Indian	Ind
Lacerta	Lacertae	The Lizard	Lac
Leo	Leonis	The Lion	Leo
Leo Minor	Leonis Minoris	The Little Lion	LMi
Lepus	Leporis	The Hare	Lep
Libra	Librae	The Scales	Lib
Lupus	Lupi	The Wolf	Lup
Lynx	Lyncis	The Lynx	Lyn
Lyra	Lyrae	The Lyre	Lyr
Mensa	Mensae	Table Mountain	Men
Microscopium	Microscopii	The Microscope	Mic
Monoceros	Monocerotis	The Unicorn	Mon
Musca	Muscae	The Fly	Mus
Norma	Normae	The Level	Nor
Octans	Octantis	The Octant	Oct
Ophiuchus	Ophiuchi	The Serpent Bearer	Oph
Orion	Orionis	The Hunter	Ori
Pavo	Pavonis	The Peacock	Pav
Pegasus	Pegasi	The Winged Horse	Peg
Perseus	Persei	Perseus	Per
Phoenix	Phoenicis	The Phoenix	Phe
Pictor	Pictoris	The Painter's Easel	Pic
Pisces	Piscium	The Fishes	Psc

CONSTELLATIONS

NAME	GENITIVE	TRANSLATION	SHORT FORM
Piscis Austrinus	Piscis Austrini	The Southern Fish	PsA
Puppis	Puppis	The Stern	Pup
Pyxis	Pyxidis	The Compass	Pyx
Reticulum	Reticuli	The Net	Ret
Sagitta	Sagittae	The Arrow	Sge
Sagittarius	Sagittarii	The Archer	Sgr
Scorpius	Scorpii	The Scorpion	Sco
Sculptor	Sculptoris	The Sculptor	Scl
Scutum	Scuti	The Shield	Sct
Serpens	Serpentis	The Serpent	Ser
Sextans	Sextantis	The Sextant	Sex

CONSTELLATIONS

NAME	GENITIVE	TRANSLATION	SHORT FORM
Taurus	Tauri	The Bull	Tau
Telescopium	Telescopii	The Telescope	Tel
Triangulum	Trianguli	The Triangle	Tri
Tri. Australe	Trianguli Australis	The Southern Triangle	TrA
Tucana	Tucanae	The Toucan	Tuc
Ursa Major	Ursae Majoris	The Great Bear	UMa
Ursa Minor	Ursae Minoris	The Little Bear	UMi
Vela	Velae	The Sail	Vel
Virgo	Virginis	The Virgin	Vir
Volans	Volantis	The Flying Fish	Vol
Vulpecula	Vulpeculae	The Fox	Vul

THE BRIGHTEST STARS

NAME	CONSTELLATION	MAGNITUDE	DISTANCE	LUMINOSITY	TYPE
Sirius	Canis Major	-1.4	8.6 light-years	23 Suns	White main sequence
Canopus	Carina	-0.7	300 light-years	13,000 Suns	Yellow supergiant
Alpha Centauri	Centaurus	-0.3	4.3 light-years	1.9 Suns	Yellow main sequence
Arcturus	Boötes	0.0	37 light-years	70 Suns	Orange giant
Vega	Lyra	0.0	25 light-years	48 Suns	White main sequence
Capella	Auriga	0.1	42 light-years	57 Suns	Yellow giant
Rigel	Orion	0.1	860 light-years	44,000 Suns	White supergiant
Procyon	Canis Minor	0.4	11.4 light-years	7.5 Suns	White main sequence
Achernar	Eridanus	0.5	91 light-years	1,000 Suns	White main sequence
Betelgeuse	Orion	0.5 average	430 light-years	9,100 Suns	Red supergiant
Beta Centauri	Centaurus	0.6	525 light-years	12,000 Suns	Blue main sequence
Acrux	Crux	0.8	320 light-years	4,000 Suns	Blue main sequence
Altair	Aquila	0.8	17 light-years	10 Suns	White main sequence
Aldebaran	Taurus	0.8	65 light-years	145 Suns	Red giant
Antares	Scorpius	1.0 average	605 light-years	11,000 Suns	Red supergiant
Spica	Virgo	1.0	260 light-years	2,100 Suns	Blue-white main sequence
Pollux	Gemini	1.1	34 light-years	30 Suns	Orange giant
Fomalhaut	Piscis Austrinus	1.2	25 light-years	16 Suns	White main sequence
Becrux	Crux	1.3 average	460 light-years	3,000 Suns	Blue main sequence
Deneb	Cygnus	1.3	3,000 light-years	250,000 Suns	White supergiant

Spaceprobe missions

SUN

NAME	COUNTRY	LAUNCH DATE	ARRIVAL/DURATION	COMMENTS
Pioneer 5	United States	March 11, 1960	106 days	First probe in orbit around the Sun
Helios 1	United States/Germany	December 10, 1974	Mid-1975 (flyby)	First close approach to Sun – studied solar wind
Solar Maximum Mission	United States	February 14, 1980	9 years	Earth-orbiting satellite to study solar flares
Ulysses	United States/Europe	October 6, 1990	July 1994 (first flyby)	Flew over the Sun's north and south poles
SOHO	United States/Europe	December 2, 1995	Still operating in 2001	Ultraviolet observatory in orbit around the Sun
TRACE	United States	April 2, 1998	Still operating in 2001	Transition Region and Coronal Explorer

MERCURY

NAME	COUNTRY	LAUNCH DATE	ARRIVAL/DURATION	COMMENTS
Mariner 10	United States	November 3, 1973	March 1974 (first flyby)	Visited Venus before three Mercury flybys

VENUS

NAME	COUNTRY	LAUNCH DATE	ARRIVAL/DURATION	COMMENTS
Mariner 2	United States	August 27, 1962	December 14, 1962 (flyby)	First successful Venus flyby
Venera 7	Soviet Union	August 17, 1970	December 15, 1970	First successful landing on Venusian surface
Pioneer Venus 1	United States	May 20, 1978	December 4, 1978 (13 years)	First radar mapping of the surface from orbit
Venera 13	Soviet Union	October 30, 1981	March 1, 1982	First soil analysis of surface
Vega 1	Soviet Union	December 15, 1984	June 9, 1985 (flyby)	Lander and weather balloon deployed
Magellan	United States	May 4, 1989	August 10, 1990 (3 years)	Comprehensive radar mapping of the planet

MOON

NAME	COUNTRY	LAUNCH DATE	ARRIVAL/DURATION	COMMENTS
Lunik 2	Soviet Union	September 12, 1959	September 14, 1959	First successful lunar impact
Lunik 3	Soviet Union	October 4, 1959	October 7, 1959	First images returned from the lunar farside
Luna 9	Soviet Union	January 31, 1966	February 4, 1966	First soft landing and pictures from surface
Luna 10	Soviet Union	March 31, 1966	April 3, 1966 (8 weeks)	First spacecraft in lunar orbit
Surveyor 1	United States	May 30, 1966	June 2, 1966 (8 months)	First successful US landing and image return
Lunar Orbiter 1	United States	August 10, 1966	August 14, 1966 (11 weeks)	Photographic survey from orbit, followed by impact
Surveyor 3	United States	April 17, 1967	April 20, 1967 (2 weeks)	Soft landing and analysis of the lunar soil
Luna 16	Soviet Union	September 12, 1970	12 days to return sample	Landing and return of rock samples to Earth
Clementine	United States	January 25, 1994	2 months	Lunar surface mapping from orbit
Lunar Prospector	United States	January 7, 1998	18 months	Lunar mineral and water prospecting from orbit

MARS

NAME	COUNTRY	LAUNCH DATE	ARRIVAL/DURATION	COMMENTS
Mariner 4	United States	November 28, 1964	July 14, 1965 (flyby)	First successful flyby and return of photos
Mars 3	Soviet Union	May 28, 1971	December 2, 1971	Brief successful landing on Martian surface
Mariner 9	United States	May 30, 1971	November 13, 1971 (349 days)	First probe in Martian orbit
Viking 1	United States	August 20, 1975	July 4, 1976 (6 years)	Successful landing and soil/atmosphere analysis
Mars Global Surveyor	United States	November 7, 1996	September 11,1997	Comprehensive photographic mapping from orbit
Mars Pathfinder	United States	December 4, 1996	July 4, 1997 (83 days)	Landing and deployment of Sojourner mini-rover
2001 Mars Odyssey	United States	April 7, 2001	October 24, 2001	Geological survey from orbit

ASTEROIDS AND COMETS

NAME	COUNTRY	LAUNCH DATE	ARRIVAL/DURATION	COMMENTS
ICE	United States	August 12, 1978	September 11, 1985 (flyby)	First comet flyby (Comet Giaccobini-Zinner)
Vega 1	Soviet Union	December 15, 1985	March 6, 1986 (flyby)	Venus/Comet Halley flyby
Suisei	Japan	August 18, 1985	March 8, 1986 (flyby)	Japanese Comet Halley flyby
Giotto	Europe	July 2, 1985	March 14, 1986 (flyby)	Successful flight through coma of Comet Halley
NEAR	United States	February 17, 1996	February 14, 2000 (1 year)	Exploration of Near-Earth Asteroid Eros
Deep Space 1	United States	October 24, 1998	July 28, 1999 (flyby)	Asteroid flyby using revolutionary ion engine

JUPITER

NAME	COUNTRY	LAUNCH DATE	ARRIVAL/DURATION	COMMENTS
Pioneer 10	United States	March 2, 1972	December 4, 1973 (flyby)	First Jupiter flyby and image return
Pioneer 11	United States	April 5, 1973	December 5, 1974 (flyby)	Used Jupiter for "gravitational slingshot" to Saturn
Voyager 1	United States	September 5, 1977	March 5, 1979 (flyby)	First "Grand Tour" mission – discovered ring
Voyager 2	United States	August 20, 1977	July 9, 1979 (flyby)	Discovered volcanic activity on Io
Galileo	United States	October 18, 1989	December 7, 1995	Detailed study of planet and moons from Jupiter orbit

SATURN

NAME	COUNTRY	LAUNCH DATE	ARRIVAL/DURATION	COMMENTS
Pioneer 11	United States	April 5, 1973	September 1, 1979 (flyby)	First Saturn flyby – reconnaissance for Voyagers
Voyager 1	United States	September 5, 1977	November 12, 1980 (flyby)	First flyby of Titan
Voyager 2	United States	August 20, 1977	August 25, 1981 (flyby)	Obtained 16,000 images of Saturn and its moons
Cassini	United States	October 15, 1997	July 1, 2004	Will enter orbit and deploy a probe to land on Titan

URANUS

NAME	COUNTRY	LAUNCH DATE	ARRIVAL/DURATION	COMMENTS
Voyager 2	United States	August 20, 1977	January 24, 1986 (flyby)	First flyby of Uranus and moons

NEPTUNE

NAME	COUNTRY	LAUNCH DATE	ARRIVAL/DURATION	COMMENTS
Voyager 2	United States	August 20, 1977	August 25, 1989 (flyby)	First flyby of Neptune – discovered activity on Triton

Satellites of the Solar System

EARTH'S MOON

NAME	MIN DISTANCE	MAX DISTANCE	ORBITAL PERIOD	DIAMETER
The Moon	365,180 km (226,820 miles)	403,620 km (250,696 miles)	27 days, 8 hours	3,476 km (2,159 miles)

MOONS OF MARS

NAME	MIN DISTANCE	MAX DISTANCE	ORBITAL PERIOD	DIAMETER
Phobos	9,240 km (5,740 miles)	9,520 km (5,910 miles)	7 hours, 40 minutes	22 km (14 miles)
Deimos	23,460 km (14,570 miles)	23,470 km (14,580 miles)	30 hours, 17 minutes	13 km (8 miles)

MOONS OF JUPITER

NAME	MIN DISTANCE	MAX DISTANCE	ORBITAL PERIOD	DIAMETER
Metis	Circular orbit of 127,960 km (79,480 miles)		7 hours, 6 minutes	46 km (29 miles)
Adrastea	Circular orbit of 128,980 km (80,110 miles)		7 hours, 9 minutes	120 km (75 miles)
Amalthea	Circular orbit of 181,300 km (112,610 miles)		11 hours, 57 minutes	200 km (124 miles)
Thebe	217,460 km (135,070 miles)	226,340 km (140,580 miles)	16 hours, 12 minutes	100 km (62 miles)
Io	404,740 km (251,390 miles)	438,460 km (272,340 miles)	42 hours, 28 minutes	3,642 km (2,262 miles)
Europa	664,190 km (412,540 miles)	677,610 km (420,880 miles)	3 days, 12.5 hours	3,130 km (1,944 miles)
Ganymede	Circular orbit of 1,070,000 km (664,600 miles)		7 days, 4 hours	5,268 km (3,272 miles)
Callisto	1,864,170 km (1,157,870 miles)	1,901,830 km (1,181,260 miles)	16 days, 16.5 hours	4,806 km (2,985 miles)
S/1975 J1	5,948,000 km (3,694,410 miles)	8,922,000 km (5,541,620 miles)	130 days	8 km (5 miles)
Leda	9,429,900 km (5,857,080 miles)	12,758,100 km (7,924,290 miles)	239 days	10 km (6 miles)
Himalia	9,643,200 km (5,989,570 miles)	13,316,800 km (8,271,300 miles)	251 days	170 km (106 miles)
Lysithea	10,430,800 km (6,478,760 miles)	13,009,200 km (8,080,250 miles)	259 days	24 km (15 miles)
Elara	9,272,230 km (5,759,150 miles)	14,201,770 km (8,820,980 miles)	260 days	80 km (50 miles)
S/2000 J11	9,870,120 km (6,130,510 miles)	15,437,880 km (9,588,750 miles)	290 days	4 km (2 miles)
S/2000 J10	17,115,000 km (10,630,440 miles)	23,635,000 km (14,680,120 miles)	591 days	4 km (2 miles)
S/2000 J3	15,135,090 km (9,400,680 miles)	26,330,910 km (16,354,600 miles)	605 days	5 km (3 miles)
S/2000 J5	16,815,200 km (10,444,220 miles)	25,222,800 km (15,666,340 miles)	618 days	5 km (3 miles)
S/2000 J7	17,987,700 km (11,172,480 miles)	24,336,300 km (15,115,710 miles)	626 days	7 km (4 miles)
Ananke	17,596,000 km (10,929,190 miles)	24,804,000 km (15,406,210 miles)	631 days	30 km (19 miles)
S/2000 J9	16,300,500 km (10,124,530 miles)	27,167,500 km (16,874,220 miles)	652 days	5 km (3 miles)
S/2000 J4	14,266,200 km (8,860,990 miles)	29,629,800 km (18,403,600 miles)	661 days	3 km (2 miles)
Carme	17,854,000 km (11,089,440 miles)	27,346,000 km (16,985,090 miles)	692 days	40 km (25 miles)
S/2000 J6	16,420,320 km (10,198,960 miles)	29,191,680 km (18,131,480 miles)	704 days	4 km (2 miles)
Pasiphae	14,570,000 km (9,049,690 miles)	32,430,000 km (20,142,860 miles)	735 days	50 km (31 miles)
S/2000 J8	11,054,870 km (6,866,380 miles)	35,987,130 km (22,352,260 miles)	733 days	5 km (3 miles)
Sinope	17,064,000 km (10,598,760 miles)	30,336,000 km (18,842,240 miles)	758 days	36 km (22 miles)
S/2000 J2	16,431,520 km (10,205,910 miles)	31,896,480 km (19,811,480 miles)	766 days	5 km (3 miles)
S/2000 J1	21,137,520 km (13,128,890 miles)	27,454,480 km (17,052,470 miles)	774 days	5 km (3 miles)

MOONS OF SATURN

NAME	MIN DISTANCE	MAX DISTANCE	ORBITAL PERIOD	DIAMETER
Pan	Circular orbit of 133,583 km (82,971 miles)		13 hours, 48 minutes	20 km (12 miles)
Atlas	137,230 km (85,230 miles)	138,580 km (85,750 miles)	14 hours, 27 minutes	32 km (20 miles)
Prometheus	139,070 km (86,380 miles)	139,630 km (86,730 miles)	14 hours, 43 minutes	108 km (67 miles)
Pandora	141,130 km (87,660 miles)	142,270 km (88,360 miles)	15 hours, 6 minutes	86 km (53 miles)
Epimetheus	150,060 km (93,200 miles)	152,790 km (94,900 miles)	16 hours, 41 minutes	122 km (76 miles)
Janus	150,410 km (93,420 miles)	152,530 km (94,740 miles)	16 hours, 41 minutes	174 km (108 miles)
Mimas	181,810 km (112,930 miles)	189,230 km (117,530 miles)	22 hours, 37 minutes	398 km (247 miles)
Enceladus	237,070 km (147,250 miles)	238,970 km (148,430 miles)	32 hours, 53 minutes	498 km (309 miles)
Calypso	Circular orbit of 294,660 km (183,020 miles)		45 hours, 19 minutes	24 km (15 miles)
Telesto	Circular orbit of 294,660 km (183,020 miles)		45 hours, 19 minutes	24 km (15 miles)
Tethys	Circular orbit of 294,660 km (183,020 miles)		45 hours, 19 minutes	1,058 km (657 miles)
Dione	Circular orbit of 377,400 km (234,410 miles)		2 days, 17.5 hours	1,120 km (696 miles)
Helene	373,630 km (232,070 miles)	381,170 km (236,750 miles)	2 days, 17.5 hours	32 km (20 miles)
Rhea	Circular orbit of 527,040 km (327,350 miles)		4 days, 12.5 hours	1,530 km (950 miles)
Titan	1,185,190 km (736,140 miles)	1,258,500 km (781,670 miles)	15 days, 22.5 hours	5,150 km (3,199 miles)
Hyperion	1,332,990 km (827,940 miles)	1,629,210 km (1,011,930 miles)	21 days, 7 hours	310 km (193 miles)
Iapetus	3,454,460 km (2,145,630 miles)	3,668,140 km (2,278,350 miles)	79 days, 8 hours	1,440 km (894 miles)
S/2000 S5	9,504,850 km (5,903,630 miles)	13,234,030 km (8,219,890 miles)	453 days	17 km (11 miles)
S/2000 S6	7,208,240 km (4,477,160 miles)	15,530,650 km (9,646,370 miles)	453 days	14 km (9 miles)
Phoebe	10,685,400 km (6,636,890 miles)	15,218,600 km (9,452,550 miles)	550 days	220 km (137 miles)
S/2000 S2	8,048,370 km (4,998,990 miles)	21,871,210 km (13,584,600 miles)	683 days	25 km (16 miles)
S/2000 S8	12,157,370 km (7,551,160 miles)	18,659,790 km (11,589,930 miles)	712 days	8 km (5 miles)
S/2000 S3	11,457,700 km (7,116,580 miles)	21,753,030 km (13,511,200 miles)	804 days	45 km (28 miles)
S/2000 S11	11,019,530 km (6,844,430 miles)	24,584,760 km (15,270,040 miles)	887 days	30 km (19 miles)
S/2000 S12	16,146,550 km (10,028,910 miles)	19,457,750 km (12,085,560 miles)	880 days	7 km (4 miles)
S/2000 S4	6,534,440 km (4,058,660 miles)	29,369,050 km (18,213,770 miles)	899 days	16 km (10 miles)
S/2000 S10	6,878,510 km (4,272,370 miles)	29,324,170 km (18,213,770 miles)	906 days	10 km (6 miles)
S/2000 S9	13,432,390 km (8,343,100 miles)	23,368,680 km (14,514,710 miles)	928 days	7 km (4 miles)
S/2000 S7	9,969,200 km (6,192,050 miles)	30,721,420 km (19,081,260 miles)	1,077 days	7 km (4 miles)
S/2000 S1	14,352,420 km (8,914,550 miles)	32,322,120 km (20,075,850 miles)	1,326 days	20 km (12 miles)

Satellites of the Solar System

MOONS OF URANUS

NAME	MIN DISTANCE	MAX DISTANCE	ORBITAL PERIOD	DIAMETER
Cordelia	Circular orbit of 49,750 km (30,900 miles)		8 hours, 2 minutes	26 km (16 miles)
Ophelia	53,230 km (33,060 miles)	54,300 km (33,730 miles)	9 hours, 2 minutes	32 km (20 miles)
Bianca	59,110 km (36,710 miles)	59,230 km (36,790 miles)	10 hours, 26 minutes	44 km (27 miles)
Cressida	Circular orbit of 61,770 km (38,370 miles)		11 hours, 8 minutes	66 km (41 miles)
Desdemona	Circular orbit of 62,660 km (38,920 miles)		11 hours, 23 minutes	58 km (36 miles)
Juliet	64,290 km (39,930 miles)	64,420 km (40,010 miles)	11 hours, 50 minutes	84 km (52 miles)
Portia	Circular orbit of 66,100 km (41,050 miles)		12 hours, 19 minutes	110 km (68 miles)
Rosalind	Circular orbit of 69,930 km (43,430 miles)		13 hours, 9 minutes	58 km (36 miles)
Belinda	Circular orbit of 75,260 km (46,740 miles)		14 hours, 59 minutes	68 km (42 miles)
S/1986U10	Tentative circular orbit of 75,260 km (46,740 miles)		14 hours, 59 minutes	40 km (25 miles)
Puck	Circular orbit of 86,000 km (53,420 miles)		18 hours, 17 minutes	154 km (96 miles)
Miranda	129,410 km (80,380 miles)	130,190 km (80,860 miles)	33 hours, 55 minutes	474 km (294 miles)
Ariel	190,670 km (118,430 miles)	191,810 km (119,140 miles)	2 days, 12.5 hours	1,160 km (720 miles)
Umbriel	264,670 km (164,390 miles)	267,330 km (166,040 miles)	4 days, 3.5 hours	1,170 km (727 miles)
Titania	434,970 km (270,170 miles)	436,710 km (271,250 miles)	8 days, 17 hours	1,580 km (981 miles)
Oberon	582,020 km (361,500 miles)	583,180 km (362,220 miles)	13 days, 11 hours	1,520 km (944 miles)
Caliban	6,599,160 km (4,098,850 miles)	7,762,440 km (4,821,390 miles)	580 days	60 km (37 miles)
Stephano	6,779,120 km (4,210,640 miles)	9,078,480 km (5,638,800 miles)	672 days	20 km (12 miles)
Sycorax	6,010,330 km (3,733,120 miles)	18,224,870 km (11,319,800 miles)	1,289 days	120 km (75 miles)
Prospero	11,124,260 km (6,909,480 miles)	21,787,740 km (13,532,760 miles)	2,009 days	30 km (19 miles)
Setebos	8,509,550 km (5,285,430 miles)	27,095,250 km (16,829,350 miles)	2,283 days	30 km (19 miles)

MOONS OF NEPTUNE

NAME	MIN DISTANCE	MAX DISTANCE	ORBITAL PERIOD	DIAMETER
Naiad	48,210 km (29,950 miles)	48,240 km (29,960 miles)	7 hours, 4 minutes	58 km (36 miles)
Thalassa	50,070 km (31,100 miles)	50,090 km (31,110 miles)	7 hours, 29 minutes	80 km (50 miles)
Despina	52,520 km (32,620 miles)	52,530 km (32,630 miles)	8 hours, 2 minutes	150 km (93 miles)
Galatea	61,950 km (38,480 miles)	61,960 km (38,480 miles)	10 hours, 18 minutes	60 km (99 miles)
Larissa	73,450 km (45,620 miles)	73,650 km (45,750 miles)	13 hours, 19 minutes	192 km (119 miles)
Proteus	117,600 km (73,040 miles)	17,690 km (73,100 miles)	26 hours, 56 minutes	418 km (260 miles)
Triton	Circular orbit of 354,760 km (220,350 miles)		5 days, 21 hours	2,706 km (1,681 miles)
Nereid	1,361,810 km (845,850 miles)	9,664,990 km (6,003,100 miles)	360 days	340 km (211 miles)

PLUTO'S MOON

NAME	MIN DISTANCE	MAX DISTANCE	ORBITAL PERIOD	DIAMETER
Charon	19,440 km (12,070 miles)	19,830 km (12,320 miles)	6 days, 9 hours	1,250 km (776 miles)

Useful addresses and websites

AMERICAN ASSOCIATION OF VARIABLE STAR OBSERVERS
25 Birch Street, Cambridge, MA 02138, USA
www.aavso.org

AMERICAN ASTRONOMICAL SOCIETY
2000 Florida Avenue NW, Suite 400,
Washington D.C. 20009, USA
www.aas.org
Professional organization with a large program of public events

ASSOCIATION OF LUNAR AND PLANETARY OBSERVERS
P.O. Box 171302, Memphis, TN 38187, USA
www.lpl.arizona.edu/alpo/

ASTRONOMICAL SOCIETY OF THE PACIFIC
390 Ashton Avenue, San Francisco, CA 94112, USA
www.aspsky.org
International society for professional and amateur astronomers

AUSTRALIAN ASTRONOMICAL SOCIETIES
A variety of local amateur societies exist – see the list at:
www.atnf.csiro.au

BRITISH ASTRONOMICAL ASSOCIATION
Burlington House, Piccadilly, London W1V 9AG, UK
www.ast.cam.ac.uk/~baa

INTERNATIONAL DARK-SKY ASSOCIATION
3225 N. First Avenue, Tucson, AZ 85719, USA
www.darksky.org
Campaigning to preserve dark skies from light pollution

INTERNATIONAL METEOR ORGANIZATION
161 Vance Street, Chula Vista, CA 91910, USA
www.imo.net
Coordinates international observing campaigns for major
meteor showers and storms

THE PLANETARY SOCIETY
65 N. Catalina Avenue, Pasadena, CA 91106, USA
www.planetary.org
Promoting the exploration of the Solar System

ROYAL ASTRONOMICAL SOCIETY OF CANADA
136 Dupont Street, Toronto, Ontario, M5R 1V2, Canada
www.rasc.ca

ROYAL ASTRONOMICAL SOCIETY OF NEW ZEALAND
P.O. Box 3181, Wellington, New Zealand
www.rasnz.org.nz

ASTRONOMY.COM
www.astronomy.com/home.asp
News and information from *Astronomy* magazine

ASTROWEB
www.cv/nrao.edu/fits/www/astroweb.html
A daily updated list of links to astronomy websites

CENTRAL BUREAU FOR ASTRONOMICAL TELEGRAMS
cfa-www.harvard.edu/cfa/ps/cbat.html
Reporting centre for astronomical discoveries

EUROPEAN SPACE AGENCY
sci.esa.int

EXTRASOLAR PLANETS ENCYCLOPEDIA
www.obspm.fr/encycl/encycl.html

HEAVENS-ABOVE!
www.heavens-above.com
Locator maps for planets, asteroids, satellites and more

HUBBLE SPACE TELESCOPE
hubble.stsci.edu

JUPSAT95
indigo.ie/~gnugent/JupSat95/
Shareware program plots the Galilean moons on your PC

NASA
www.nasa.gov
Information on US manned spaceflights, probes and more

SEARCH FOR EXTRATERRESTRIAL INTELLIGENCE
www.seti.org

SPACE WEATHER
www.spaceweather.com
Aurora predictions, updates on solar flares and more

SKY & TELESCOPE MAGAZINE
www.skypub.com/skytel/skytel.shtml

STUDENTS FOR EXPLORATION AND DEVELOPMENT OF SPACE
www.seds.org
Includes comprehensive Solar System and Messier object tours

JET PROPULSION LABORATORY
www.jpl.nasa.gov
Links to many of NASA's missions

Glossary

ACCRETION DISC
A disc of spiralling gas that forms as a star pulls in material from around it, or from a neighbouring star. Material does not fall directly onto the star because it is already moving in its own circular orbit.

ANTISOLAR POINT
The point in the sky directly opposite the Sun. At night, the antisolar point may glow faintly as dust in the plane of the Solar System reflects sunlight – a phenomenon called the gegenschein.

ASTEROID
A small, rocky body too small to be classified as a planet, most often found in orbits between Jupiter and Earth. With a few exceptions, asteroids are too small to pull themselves into spherical shapes.

ASTRONOMICAL UNIT (AU)
A standard measure of astronomical distance defined as the average radius of Earth's orbit – 92.9 million miles (149.6 km).

BAILY'S BEADS
A chain of bright points of light seen at the beginning or end of a total solar eclipse, caused by light from the Sun shining through valleys on the Moon's surface.

BIG BANG
The enormous explosion that created the Universe, approximately 12 billion years ago.

BINARY STAR
A pair of stars locked in orbit around one another. Binary stars can orbit each other with periods ranging from a few hours to a thousand years or more.

BLACK HOLE
Any object so dense that its gravity prevents light from escaping it. Black holes range from the collapsed cores of massive stars to objects millions of times heavier at the centres of galaxies.

BLUESHIFT
A Doppler shift of an object's light toward the blue (high-frequency) end of the spectrum, indicating the source of the light is moving toward us.

BROWN DWARF
A small, faint object that forms in the same way as a star, but does not have enough mass to begin shining by hydrogen-burning nuclear reactions.

EXTREME NUMBERS
Astronomy is full of extremely big and small numbers. In order to write such numbers quickly and easily, scientists use a special notation system, based on powers of 10. A power is shown by 10 followed by a superscripted number, indicating how many times it should be multiplied by itself – for example, 10^2 indicates 10 x 10, or 100, and 10^9 indicates 10 to the power of 9, or 1,000,000,000 (a billion). Small numbers are indicated by negative powers, indicating that the number is 1 divided by 10 to a certain power. So 10^{-1} is 0.1, 10^{-3} is 0.001 (one thousandth), and so on.

CELESTIAL EQUATOR
An imaginary line around the sky marking the extension of Earth's equator into space. The celestial equator lies midway between the celestial poles, and marks 0 degrees declination, the division between the sky's northern and southern hemispheres.

CELESTIAL POLES
The points in the sky directly above Earth's north and south poles, around which constellations appear to rotate.

CENTRE OF MASS
The balance point of a system containing two or more objects, around which every object in the system orbits.

CEPHEID
An unstable giant star that varies in size and brightness in a repeating cycle, named for the prototype star Delta Cephei. The period of a Cepheid's oscillations is directly linked to its true brightness so, by comparing this to the apparent brightness from Earth, astronomers can estimate their distances.

CHANDRASEKHAR LIMIT
An upper limit to the mass of a white dwarf – the collapsed core of a dead Sunlike star. Above 1.4 solar masses, the white dwarf's gravity is so great that its atoms collapse and it shrinks to a tiny neutron star.

CIRCUMPOLAR
For an observer at a particular latitude, any star or constellation that circles the celestial pole, but never sets.

COMET
A minor member of the Solar System made mostly of ice. Most comets lie beyond the orbit of Neptune, but some fall into elliptical orbits that bring them close to the Sun. Material evaporating from the surface then forms a vast gaseous coma around the solid nucleus, and may stream away in a long tail.

CONJUNCTION
An appearance of two astronomical objects close to each other in the sky.

CORONA
A layer of hot, thin gas extending above the visible surface of the Sun. Intense magnetism here can heat the corona to 3 to 4 million °F (1 to 2 million °C).

DARK MATTER
Material that makes up a large proportion of the Universe's mass, but is invisible because it is not luminous and is not illuminated by another light source. Dark matter can only be detected by the effects of its gravity.

DAY
The time a planet takes to turn on its axis. Earth's day is 23 hours, 56 minutes long – rounded up to 24 hours for convenience of timekeeping.

DECLINATION
A measure of the position of a star relative to the celestial equator, measured in degrees above or below it. The north celestial pole has declination +90 degrees, and the south celestial pole has declination -90 degrees.

DECOUPLING
The moment 300,000 years after the Big Bang when the Universe became transparent. Decoupling represents the earliest event that we can observe.

DEGREE
A measure of size and angle in the night sky. One degree is 1/360th of the circle around the whole celestial sphere.

DEUTERIUM
A form of hydrogen that contains a neutron in addition to its proton and electron. The neutron has no effect on the atom's chemistry, but doubles its mass. Deuterium is formed as an intermediate step in the nuclear fusion of hydrogen into helium.

DIFFRACTION
The spreading out or dispersal of light waves or any other waves as they pass an edge or go through a slit. Diffraction affects light of different colours by different amounts, and can be used to create a spectrum.

DOPPLER SHIFT
A change in the frequency of waves such as light waves passing an observer, caused when the source of the waves is moving toward or away from the observer.

ECLIPTIC
The path taken by the Sun around Earth's sky each year. The ecliptic represents the plane of Earth's own orbit, and most other planets of the Solar System orbit on or near to it.

EDGE-ON
The view of a flat disc, such as a spiral galaxy or an accretion disc, in which we see only the thin edge of the disc.

ELECTRON
A lightweight subatomic particle with little mass and a single, negative, electrical charge. An atom consists of electrons orbiting a nucleus that contains most of its mass.

ELLIPTICAL ORBIT
The most common type of orbit. An ellipse is a type of stretched circle with two foci, or focuses. The system's centre of mass lies at one of the foci. An ellipse's shape is measured by its eccentricity – a circular orbit is a special type of ellipse with zero eccentricity and the two foci in the same place.

EMISSION NEBULA
A gas cloud in space that emits its own light. The emission is usually caused by intense heating – for example, by young stars embedded in the nebula.

EQUATORIAL COORDINATES
A widely used astronomical coordinate system that measures positions by declination above or below the celestial equator and right ascension relative to the First Point of Aries.

EVENT HORIZON
The boundary around a black hole, marking the point where its gravity becomes so strong that not even light can escape it.

FACE-ON
The view of a flat disc (such as a spiral galaxy or accretion disc) in which the upper or lower surface of the disc is fully displayed to an Earthbound observer.

FESTOONS
Diagonal streams of cloud linking cloud belts or zones in the atmosphere of a gas-giant planet.

FILAMENTS
The largest known structure in the Universe, made up of bubbles, sheets and chains of millions of galaxies.

FIRST POINT OF ARIES
The point where the Sun's path around the sky (the ecliptic) passes from the southern to the northern hemisphere of the sky at the beginning of northern spring. The First Point of Aries, now carried into the constellation of Pisces by precession, is defined as zero hours of right ascension in the equatorial coordinate system.

GALAXY
A large system containing anything from millions to hundreds of billions of stars, isolated from its neighbours in space by hundreds of thousands of light-years. Galaxies can have elliptical, spiral, or irregular structures.

GAMMA RAYS
The most powerful form of radiation in the Universe, with extremely short wavelengths and high energies. Gamma rays are produced only by violent astronomical objects.

GAS-GIANT PLANET
A planet such as Jupiter and Saturn consisting of a huge gas envelope around a roughly Earth-sized rocky core.

GENERAL RELATIVITY
The theory proposed by Albert Einstein that large masses such as planets and stars distort space and time around them.

GLOBULAR CLUSTER
A dense ball containing thousands of ancient, yellow and red stars, in orbit around a galaxy.

HELIUM
The second-lightest element in the Universe, produced by the nuclear fusion of hydrogen in stars and in the Big Bang. A helium atom consists of a nucleus of two protons and two neutrons, orbited by a pair of electrons.

HERBIG-HARO OBJECT
A nebula formed by a young star that is undergoing violent fluctuations and flinging off material as it settles onto the main sequence.

HERTZSPRUNG-RUSSELL DIAGRAM
A diagram comparing the luminosities of stars with their colours. The H-R diagram is the key to understanding stellar evolution.

HUBBLE'S LAW
A law resulting from the general expansion of the Universe – the farther away a distant galaxy is from Earth, the faster it is receding and the larger the redshift of its light.

HYDROGEN
The simplest and most plentiful element in the Universe, made up of a single proton and a single electron. Hydrogen is also the major power source for the stars.

IC NUMBER
An object's designation in the Index Catalogue – an extension to the New General Catalogue (NGC) listing 5,386 star clusters, galaxies and nebulae.

INFLATION
The sudden expansion of a small portion of the Universe a fraction of a second after the Big Bang, which created nearly all the matter in the Universe today.

INFRARED
Heat or radiation with lower frequencies and less energy than visible light, emitted by all warm objects.

KIRKWOOD GAPS
Empty rings within the asteroid belt, marking the orbits where asteroids have been ejected by frequent close encounters with Jupiter's powerful gravity.

KUIPER BELT
A belt of icy comets and small frozen worlds beyond the orbit of Neptune, including the ninth planet Pluto.

LIGHT-YEAR
The distance travelled by light across space in one year, approximately 5.9 million million miles.

LOCAL GROUP
The small cluster of about 30 galaxies centred on the Milky Way and the Andromeda Galaxy.

LUMINOSITY
The true brightness of a star, galaxy or other object, as opposed to its apparent magnitude as seen from Earth.

MACHO

A Massive Compact Halo Object – a faint, dark object such as a brown dwarf or black hole, which may orbit in the spherical halo around a galaxy and account for much of its unobserved dark matter.

MAGELLANIC CLOUDS

Two small irregular galaxies locked in orbit around our own Milky Way Galaxy.

MAGNETAR

A neutron star with an intense magnetic field that occasionally ruptures its surface, releasing a burst of gamma rays.

MAGNITUDE

A measure of the brightness of an astronomical object. The lower the magnitude, the brighter the object is. Naked-eye stars go down to magnitude 6.0, and the brightest stars and planets have magnitudes below zero. A difference of 1 magnitude corresponds to a change in brightness by a factor of roughly 2.5.

MAIN SEQUENCE

A line joining faint red stars to bright blue ones on the Hertzsprung-Russell diagram, where most stars are found. The main sequence represents the majority of any star's lifetime, when it is shining by burning hydrogen in its core.

MEGAPARSEC

A million parsecs, or 3.26 million light-years, a unit often used in measuring the distances of far-away galaxies.

MESSIER OBJECT

Any of 110 objects listed in the Messier catalogue. These include nearly all of the brightest star clusters, nebulae and galaxies visible from mid-northern latitudes.

MILKY WAY

The spiral galaxy of which our Sun is a member. Also the band of light across the night sky created where we look through the galaxy's dense star clouds.

MINUTE OF ARC

A measure of size and angle in the night sky. One minute of arc is $\frac{1}{60}$th of a degree, or $\frac{1}{21,600}$th the way around the whole sky.

NEBULA

A cloud of gas and/or dust in space. Nebulae may be dark, absorbing light, or glow due to emission or reflection.

NEUTRON

A subatomic particle with the same mass as a proton but no electrical charge. Neutrons and protons account for most of an atom's mass. They inhabit the nucleus, orbited by a cloud of electrons.

NEUTRON STAR

The collapsed core of a massive star left behind after a supernova. The core's strong gravity causes its atoms to collapse, forming a superdense ball of neutrons just a few miles across.

NGC NUMBER

An object's designation in the New General Catalogue of Nebulae and Star Clusters. The NGC was published in 1888 by J.L.E. Dreyer and contains 7,840 nebulae, galaxies and star clusters in all areas of the sky.

NUCLEAR FUSION

The energy source that powers the stars by combining the nuclei of lightweight elements to make heavier ones.

NUCLEUS

The central region of an atom, containing nearly all of its mass. The nucleus is made up of protons and neutrons.

OCCULTATION

An astronomical event in which one object passes in front of a more distant one. A solar eclipse is a special type of occultation.

OORT CLOUD

A huge spherical cloud of deep-frozen comets surrounding the Solar System at a distance of around a light-year.

ORBIT

The path through space taken by any object around another, under the influence of the other's gravity.

PARALLAX

The shift in the apparent positions of nearby objects against a more distant background, seen when the same scene is observed from two different points of view. Parallax is a vital tool for measuring cosmic distances.

PARSEC

The distance at which a star would show a parallax of one arcsecond when viewed from two points one Astronomical Unit apart – equivalent to 3.26 light years.

PENUMBRA

The pale border around a sunspot, between its dark central umbra and the bright surface of the Sun. Also the region of paler shadow around the path of a total solar eclipse where observers see a partial eclipse.

PHOTOSPHERE

The visible surface of the Sun or another star where its gas becomes transparent in the space of a few hundred kilometres, giving it a relatively sharp boundary.

PLANETARY NEBULA

A series of shells of gas, often with complex shapes, thrown off by a red giant as it exhausts its final fuel supplies.

PLANETESIMAL

A small world with sufficient gravity to attract other material to it. Planetesimals may eventually collide and form planets.

PRECESSION

A 25,800-year wobble in Earth's axis caused by the gravitational pull of the Sun and Moon. Over thousands of years, precession causes Earth's poles to point to different areas of the night sky.

PROMINENCE

A bright loop of material in the Sun's upper atmosphere or corona. Prominences are cooler but denser than their surroundings, and mark out the lines of the Sun's magnetic field. They appear as dark filaments when seen against the Sun's brighter disc.

PROTON

A subatomic particle with reasonably high mass and a single positive electrical charge. Protons and neutrons account for most of an atom's mass. They inhabit the nucleus, orbited by a cloud of electrons.

PROTOSTAR

A star that is still in the process of collapsing from its initial gas cloud. Protostars are not yet shining by the nuclear fusion of hydrogen – instead they glow from the heat of their gravitational contraction, and lower-energy nuclear reactions.

PULSAR

A rapidly spinning neutron star whose magnetic field creates intense beams of radiation from its magnetic poles. When the poles happen to point in our direction, we detect rapid flashes from the star.

QUASAR
A distant galaxy in the early Universe that emits brilliant light and other radiations from a tiny region at its core. Quasars are probably powered by supermassive black holes at their centres.

RADIANT
The point in the sky from which a meteor storm appears to originate, caused by the effects of perspective.

RADIO WAVES
Radiation with the lowest energies and frequencies of all, generated by most objects in the Universe.

RED DWARF
A star with less than 0.8 solar masses of material, which shines only faintly. Red dwarfs are hard to detect, but are probably very plentiful throughout our galaxy and the Universe.

RED GIANT
A Sunlike star that has exhausted the hydrogen fuel at its core and become unstable, swelling to hundreds of times its previous size, and changing colour as its surface cools.

REDSHIFT
A Doppler shift of an object's light toward the red (low-frequency) end of the spectrum, indicating that the source of light is moving away from us.

REFLECTION NEBULA
A cloud of gas in space shining by reflecting the light of nearby stars.

RIGHT ASCENSION (R.A.)
The astronomical equivalent of longitude, measured in hours, minutes and seconds east of a line from pole to pole, through the First Point of Aries.

SECOND OF ARC
A measure of size and angle in the night sky. One second of arc is $\frac{1}{60}$th of a minute of arc, $\frac{1}{3,600}$th of a degree, and $\frac{1}{1,296,000}$th the way around the whole sky.

SOLAR SYSTEM
The region of space in which the Sun's gravity is the dominant force, and all the objects such as planets, asteroids and comets contained within it.

SOLAR WIND
A stream of particles blown away from the surface of the Sun by the pressure of its radiation. The solar wind contains mostly electrons and protons, some of which become trapped by Earth's magnetic field, and create aurorae in the upper atmosphere.

SPACE-TIME
According to Einstein's relativity theory, a four-dimensional combination of space and time, in which the three dimensions of space and one of time can affect and distort each other.

SPECIAL RELATIVITY
Albert Einstein's theory explaining how objects apparently become distorted when travelling close to the speed of light.

SPECTRAL TYPE
A method of classifying stars by their colour – and therefore surface temperature – and the absorption lines in their spectrum. Stars range from extremely hot, blue O-type, through B, A, F and G (yellow stars) to orange K, and cool, red M-types.

SPECTROSCOPIC BINARY
A binary pair of stars so close together that they are only identified by Doppler shifts in their light as they orbit one another.

STELLAR WIND
A stream of particles blown out from the surface of a star, similar to the solar wind. In certain stars, stellar winds can become very powerful and cause the star to lose significant amounts of material.

SUPERGIANT
An extremely massive and luminous star. Supergiants can be of any colour, and are the end stages of the most massive stars, on their way to becoming supernovae.

SUPERNOVA
An enormous stellar explosion marking the end of a massive star's life, or a white dwarf's transformation into a neutron star.

TERRESTRIAL PLANETS
The small, rocky planets of the Solar System with relatively thin atmospheres, such as Mercury and Earth.

ULTRAVIOLET RADIATION
Radiation with slightly more energy than visible light, generated by very hot objects, such as young stars with high masses.

UMBRA
The dark region at the centre of a sunspot where the magnetic field is most intense and the lower temperatures cause gas to look dark compared with its surroundings. Also the deep shadow cast on Earth's surface by a total solar eclipse.

VAN ALLEN BELTS
A pair of doughnut-shaped radiation belts around Earth where particles from the solar wind are trapped by our planet's magnetic field.

VARIABLE STAR
A star that changes its magnitude as seen from Earth. Variability can be caused by two stars eclipsing each other, or by actual physical changes in a star.

VOIDS
Enormous, apparently empty regions of the Universe separating the filaments of galaxies. Voids may be full of dark matter.

WHITE DWARF
The collapsed core of a Sunlike star that has exhausted its fuel and blown away its outer layers. Without nuclear reactions to support it, the core collapses under its own gravity into an object roughly the size of Earth.

WIMP
A Weakly Interactive Massive Particle – theoretical particles produced in the Big Bang that are almost impossible to detect and could account for much of the Universe's dark matter.

WOLF-RAYET STAR
An extremely massive and highly luminous star. The star's brilliance creates fierce stellar winds that blow away its outer layers into space, causing it to rapidly lose mass through its lifetime.

X-RAYS
A type of radiation with very high frequencies and energies, produced by extremely hot and energetic astronomical objects, such as black holes and quasars.

YEAR
The period of time a planet takes to complete one orbit of its star.

ZODIAC
The band of 12 constellations through which the ecliptic passes, and where the planets are most frequently seen.

Celestial object index

Acknowledgments

I'd like to extend my thanks to everyone who worked on this book, but especially to Ellen Dupont, for getting the project off the ground, to Dan Green and Kim Ruderman for putting it all together, and to Robert Burnham for tireless advice. In addition, thanks to family, friends, and girlfriend for support, distraction and pizza respectively.

This book is dedicated to the ever-changing staff of the café at Border's, Oxford Street, without whom it would not have been likely.

Images:
AAO = Anglo-Australian Observatory; **BC** = Bruce Coleman; **GPL** = Galaxy Picture Library; **OSF** = Oxford Scientific Films; **RHPL** = Robert Harding Picture Library; **SPL** = Science Photo Library; **TNHM** = The Natural History Museum; **TAA** = The Art Archive
Key: b bottom **c** centre **l** left **r** right **t** top

1-2 BC; 4l OSF/Michael Sewell, 4r BC, 4.5 knocked back GPL; 5l BC, 5c BC, 5l BC; 8-9 GPL/Michael Stecker; 10-11 OSF/Michael Sewell; 12-13 background NASA/STScI; 14 GPL/Tatsuo Nakagawa; 15l SPL/Dennis Di Cicco, Peter Arnold Inc., 15r GPL/Robin Scagell; 30-32 BC; 33l SPL/John Sanford; 33r GPL/Robin Scagell; 34 BC; 35t TNHM, London, 35cl GPL/John Fletcher, 35cr TNHM, London, 35bl NASA/Carnegie Mellon University, 35br TNHM, London; 36 NASA; 37l SPL/David Parker, 37r TAA/Eileen Tweedy; 38 NASA; 39t BC, 39bl courtesy of D. Hathaway, NASA/Marshall Space Flight Center, 39br TAA/Eileen Tweedy; 40 GPL/Paul Coleman; 41t GPL /Howard Brown-Greaves, 41cl GPL/Michael Stecker, 41cr NASA, 41b eclipse map courtesy of Fred Espenak, NASA/Goddard Space Flight Center; 43tl BC, 43tr OSF/Norbert Rosing, 43b OSF/David C. Fritts; 44 OSF/NASA; 45t GPL/Robin Scagell, 45c GPL/Robin Scagell, 45b GPL/Robin Scagell; 46 OSF/Kim Westerskov; 47t BC, 47cl OSF /Kim Westerskov, 47cr GPL/Robin Scagell, 47b GPL/Howard Brown-Greaves; 48 BC; 49l GPL/John Costello, 49r Mary Evans Picture Library; 50 Lick Observatory University of California; 51 BC; 52 NASA; 53t Lick Observatory University of California, 53b TAA /British Museum; 54 NASA; 55l GPL/Robin Scagell, 55r Corbis /Historical Picture Archive; 56 NASA; 57l GPL/Robin Scagell, 57c Corbis/Roger Rossmeyer, 57r TAA/Archaeological Museum Aleppo /Dagli Orti; 58 NASA; 59 NASA; 60t NASA, 60c NASA, 60b NASA; 61tl NASA, 61tr NASA, 61b GPL/David Graham; 62l NASA, 62r NASA; 63 NASA/STScI; 64 NASA; 65t NASA, 65b NASA; 66 NASA; 67t NASA, 67tc NASA, 67bc NASA, 67b NASA; 68 NASA; 69 NASA, 69b NASA; 70t NASA, 70b NASA; 71 NASA; 72 BC; 73 GPL/Robin Scagell; 74 BC; 75tl GPL/Maurice Gavin, 75tr Corbis /Adam Woolfitt, 75b NASA; 76t NASA, 76b NASA; 77t GPL/Lowell Observatory, 77b GPL/David Jewitt; 78 GPL/Michael Stecker; 79t GPL/Paul Sutherland, 79c Digital Vision, 79b Corbis/Gianni Dagli Orti; 80-82 BC; 83tl GPL/KPNO, 83tr GPL/Robert Dalby, 83bl GPL/Robin Scagell, 83br BC; 85tl GPL/DSS, 85tr GPL/Pedro Re, 85bl Mary Evans Picture Library, 85br BC; 87l AAO/David Malin, 87r GPL/Gordon Garradd; 88l NASA, 88r GPL/Robin Scagell; 89l GPL /Robin Scagell, 89r GPL/Robin Scagell; 91l AAO/David Malin, 91c AAO/David Malin, 91r Corbis/Kevin R. Morris; 93t GPL, 93bl NASA, 93br GPL/Robin Scagell; 94 GPL/Michael Stecker; 95t NASA/STScI, 95c GPL/Robin Scagell, 95bl GPL/John Gillett, 95br RHPL/Nigel Francis; 96t BC, 96b NASA; 97t BC, 97c GPL/KPNO, 97bl GPL /Michael Stecker, 97br BC; 99 GPL/Robin Scagell; 100 BC; 101t European Southern Observatory, 101b Carnegie Institution of Washington; 102 SPL/Chris Butler; 103t Ed Grafton (website: http://ghg.net/egrafton), 103c Ed Grafton, 103b GPL/Pedro Re; 104 NASA; 105t GPL, 105ct NOAO, NSF, 105cb GPL/Gordon Garradd, 105b TAA/Museo Civico Padua/Dagli Orti; 106 BC; 107t GPL, 107bl

GPL/Maurice Gavin, 107br GPL/Adrian Catteral; 109t GPL/Michael Stecker, 109bl GPL/Robin Scagell, 109bc GPL/Robin Scagell, 109br GPL /David Ratledge; 110 AAO; 111t BC, 111bl BC, 111br AAO; 113t AAO /Royal Observatory, Edinburgh, 113c GPL/Chris Floyd, 113b GPL; 115t TAA/Royal Astronomical Society/Eileen Tweedy, 115b GPL/Robin Scagell; 116 NASA; 117t GPL/AURA/NOAO/NSF, 117bl GPL/Isaac Newton Group, 117br Malin/IAC/RGO; 118t BC, 118c GPL/STScI, 118b GPL/STScI; 119t Grove Creek Observatory, 119c GPL, 119b RHPL/Claire Leimbach; 120 AAO/David Malin; 121tl Californian Institute of Technology/Carnegie Institution of Washington, 121tr Royal Observatory, Edinburgh/David Malin, 121b GPL/Robin Scagell; 122 NASA; 123t GPL/Gordon Garrard, 123c Astronomical Images Library/DSS, 123b AAO/David Malin; 124 BC; 125t AURA/NOAO/NSF, 125bl NASA, 125br AL Kelly (website: http://www.ghg.net/akelly); 126 BC; 127t GPL; 127bl GPL/Michael Stecker, 127br GPL/Robin Scagell; 128 BC; 129t GPL/Michael Stecker, 129bl GPL /Robin Scagell, 129br GPL/David Ratledge; 130 BC; 131l GPL/Robin Scagell, 131r GPL/David Ratledge; 132l BC, 132r NASA; 133t GPL/Pedro Re, 133c GPL/N. Sharp/ AURA/NOAO/NSF, 133b TAA/Musee du Louvre Paris/Dagli Orti; 135l BC, 135r NASA; 136 NASA/2MASS Project, Umass, IPAC/Caltech, NSF; 137t Corbis/Bettmann, 137c AAO/David Malin, 137b AAO/Royal Observatory, Edinburgh; 138-140 BC; 141t AAO/Royal Observatory, Edinburgh/David Malin, 141bl BC, 141br TAA; 142 BC; 143l National Optical Astronomy Observatories, 143 r GPL/Brad Ehrhorn; 145t GPL/Michael Stecker, 145b GPL/George Sallitt; 146 GPL/N. Sharp/AURA /NOAO/NSF; 147t AAO/David Malin, 147bl Royal Observatory, Edinburgh /David Malin, 147br Corbis/Gianni Dagli Orti; 149tl GPL/Robin Scagell, 149tr AAO/Royal Observatory, Edinburgh, 149b GPL/European Southern Observatory; 150l AAO/David Malin, 150c NASA, 150r NASA; 151tl NASA, 151tr AAO/David Malin, 151b GPL/AURA /NOAO/NSF; 152 NASA; 153t GPL/Isaac Newton Telescope, 153c GPL/Pedro Re, 153b GPL /Pedro Re; 155t GPL/Nik Szymanek & Ian King, 155c GPL/AURA/NOAO/NSF, 155b GPL/Michael Stecker; 156 BC; 157l GPL/AURA/NOAO/NSF, 157r GPL/Gordon Garrard; 159t AAO/David Malin, 159bl AAO/David Malin, 159br NASA; 160-161 BC; 163 NASA; 164 GPL?; 165 BC; 167t NASA, 167b Corbis/Bettmann; 170 BC; 171-172 Science & Society Picture Library/CERN; 175-176 NASA; 178 GPL/Richard Wainscoat; 179 Corbis /Roger Rossmeyer; 180 European Southern Observatory; 181 NASA/photo courtesy of Chip Simons; 183l Corbis/Roger Ressmeyer, 183r SNO/photo courtesy of Bob Chambers; 184 GPL/Robin Scagell; 185 John Barlow; 186 GPL; 187 Meade Instruments; 188 GSFC/NASA; 189 TAA/Dagli Orti; 190 TAA/J. Enrique Molina; 191t BC, 191b TAA/Dagli Orti; 192t TAA/British Library, 192b TAA/British Library; 193t TAA/British Library, 193b TAA /Galleria degli Uffizi Florence/Dagli Orti; 194l Corbis/Bettmann, 194r Science & Society Picture Library/Science Museum; 195 Corbis/Bettmann; 196t Science & Society Picture Library/Science Museum, 196b Corbis /Hulton- Deutsch Collection; 197t GPL/Robin Scagell, 197b Popperfoto; 198l NASA, 198r NASA; 199l GPL/Carnegie Institution, 199r Corbis /Bettmann; 200 NASA; 201b NASA, 201t Bettman/Corbis; 202 NASA; 203 Corbis/Bettmann.

Illustrations:
All starmaps by Kevin Jones Associates, except those on pages 19 and 21-29 by Wil Tirion. All planet locator charts by Tim Loughhead. Other illustrations by Patrick Mulrey, Giles Sparrow, Wil Tirion, Guy Smith, Mainline Design with assistance from: Phil Gamble, Paul Montague and Ella Butler.